Photovoltaics for Sustainable Electricity and Buildings

Ali Sayigh
Editor

Photovoltaics for Sustainable Electricity and Buildings

Springer

Editor
Ali Sayigh
Chairman
World Renewable Energy Congress
Brighton, UK

ISBN 978-3-319-39278-3 ISBN 978-3-319-39280-6 (eBook)
DOI 10.1007/978-3-319-39280-6

Library of Congress Control Number: 2016959386

Printed on acid-free paper

This Springer imprint is published by Springer Nature
The registered company is Springer International Publishing AG
The registered company address is: Gewerbestrasse 11, 6330 Cham, Switzerland

Preface

In light of the role played by fossil fuels in global warming and climate change and the increasing demand for energy, renewable industries must address the need for sustainable electricity and buildings, especially within the field of photovoltaics. Presently, investment in renewable energy is double that of the investment in coal and gas according to UN sources. With this in mind, this volume brings together a number of distinguished authors to make the case for photovoltaics as one of the most effective electricity generators among all the renewable energy producers.

The book is aimed at national and regional policy makers as well as environmentalists. It will be of interest to electricity producers, researchers, building technologists, and postgraduate students in science and engineering. Photovoltaic developers and manufacturers will find this book gives them a view of the industry as it stands at the present time and where its future lies.

It is hoped that this book will encourage the building and construction industries to adopt the concept of photovoltaic cells and panels as a building material which contributes not only to a building's fabric but also to its source of electricity supply.

The book is structured to take the reader from the recent history of photovoltaics development through to modern concepts and applications.

Brighton, UK

Ali Sayigh

Contents

About the Editor

Ali Sayigh is Chairman and Founder of the World Renewable Energy Congress and Council, Director General of World Renewable Energy Network (WREN), Chairman and Founder of the Arab Solar Energy Society, and Past Chairman of the UK Solar Energy Society. Dr. Sayigh was recently elected to chair the Iraqi Energy Institute, and he actively consults on renewable and sustainable energy issues for a number of international organizations, including UNESCO, ISESCO, UNDP, ESCWA, and UNIDO. Dr. Sayigh was Director of Solar Seminars at ICTP Trieste, Italy, from 1977 to 1995; Professor of Solar Energy at King Saud, Kuwait, and Reading Universities from 1969 to 1994; and Professor of Engineering at the University of Hertfordshire from 1994 to 2004. He was the founding expert in Renewable Energy at AOPEC. He is a Fellow of the Institute of Energy, a Fellow of the Institution of Electrical Engineers, and a Chartered Engineer. He has published more than 400 papers and has contributed to and edited more than 30 books. He has been Editor and Editor-in-Chief of several international journals including Renewable Energy and the International Journal of Environmental Sciences and Technology and Editor-in-Chief of the Major Reference Work, Comprehensive Renewable Energy.

Chapter 1
Photovoltaic Technology Is Electricity for All

Ali Sayigh

Installations of photovoltaic (PV) in the 1970s and 1980s were limited to kilowatts, while in the 2010s, we speak in terms of megawatts. The cost of most renewable energy systems has been reduced by so much that they have reached parity with fossil fuels or are even cheaper. The most effective progress has been made in photovoltaic systems: the cost of turnkey installations say for 5 MW is no more than $6 million. Governments in European countries are using feed-in tariffs which has made the payback period of installing a large system less than 1.25 years. Countries such as Morocco have pledged to produce 42 % of their electricity from renewable energy by 2020, while Austria has declared that by 2050, all its energy will come from renewable sources.

It is clear from the media and UN reports that there is no country which is not utilising renewable energy to some extent, but what is urgently needed is for this use to be redoubled immediately to prevent the earth heating by more than 2 °C.

Progress in PV area of renewable energy has exceeded all expectations. Nowadays there is no country in the world without some PV applications no matter how small the country or how small the application. It is truly a revolution in manufacturing, innovation, cost reduction and efficiency improvement (Fig. 1.1).

In terms of technology, several advances have been made during the last three years, for example:

– Magnolia Solar demonstrates and has produced high-performance coated glass which consists of coating the glass with a new class of materials consisting of porous silicon dioxide nanorods. The reflection losses at the glass-air interface have been reduced from approximately 4 % to less than 1 %. At large angles

A. Sayigh (✉)
World Renewable Energy Congress, Flat 3, House 5-6, Clarendon Terrace, Brighton BN2 1FH, UK, Clarendon Terrace, Brighton 1FD E Sussex, UK
e-mail: a.sayigh@wrenuk.co.uk

A. Sayigh (ed.), *Photovoltaics for Sustainable Electricity and Buildings*,
DOI 10.1007/978-3-319-39280-6_1

Fig. 1.1 (**a**) The 80 MW Okhotnykovo Solar Park in Ukraine, as a whole, Ukraine has 13,551 MW with one array of 400 MW and plans to have 50 % of their electricity generated from PV by 2020; (**b**) Japan has added 537 MW of capacity since February 2014; (**c**) the Philippines has a renewable energy generation capacity of 5 GW

of incidence, the reflection losses have been reduced from over 25 % to less than 5 %.

– Flexible thin-film photovoltaic modules have been manufactured by Ascent Solar, the EnerPlex Kickr IV, and can be used on the beach, in picnic and on the car dashboard—it is both light and strong. Plastic cells are much cheaper and flexible. A new UK study has shown that even when using very simple and inexpensive manufacturing methods—where flexible layers of material are deposited over large areas like cling film—efficient solar cell structures can be made.

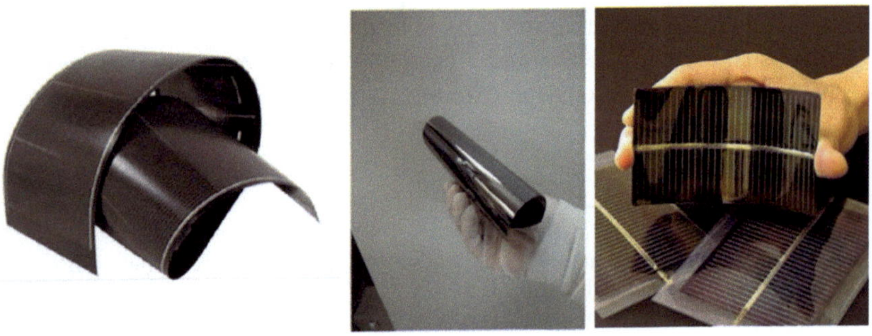

– Trina Solar has the highest efficiency record for the p-type monocrystalline silicon solar cell on an industrial Cz wafer, which integrates advanced technologies including back surface passivation and local back surface field, reaching an efficiency of 21.40 % (156 × 156 mm^2). In 2014, Trina produced 3.5 GW of PV, according to UK-based market research firm GlobalData, which represents 7.9 % of global production [1].

– Germany's Fraunhofer Institute for Solar Energy Systems ISE announced in September 2015 it has achieved a 25.1 % efficiency for a both sides contacted silicon solar cell.

– Trina Solar in December 2015 reported it has developed p-type monocrystalline silicon solar cell with an efficiency of 22.13 %. 156 × 156 mm2. This compares with 25 % the cell of 2X2 cm^2 developed by Martin Green a long time ago.

- First Solar records 17.0 % efficiency for a CdTe solar module, cell efficiency of 20.4 % in 2015.

- Boeing has set a PV efficiency world record of 37.8 % ground-based solar cell without solar concentration using multi-junction of two or three materials, reported by Global Energy World in 2013.

- The Energy Department at National Renewable Energy Lab announced in June 2013 a world record of 31.1 % conversion efficiency for a two-junction solar cell under one sun of illumination (AM 1.5, 1000 W/m^2 made of a gallium indium phosphide cell atop a gallium arsenide cell of 0.25 cm^2).
- SolenSphere Renewables produced a cell of 40 % efficiency, and when combined with the thermal energy captured by a parabolic concentrator, it approaches a total electrical and heat energy efficiency of 72 %. This is cost-effective and has cell area reduction of 1000 [2] (Figs. 1.2 and 1.3).

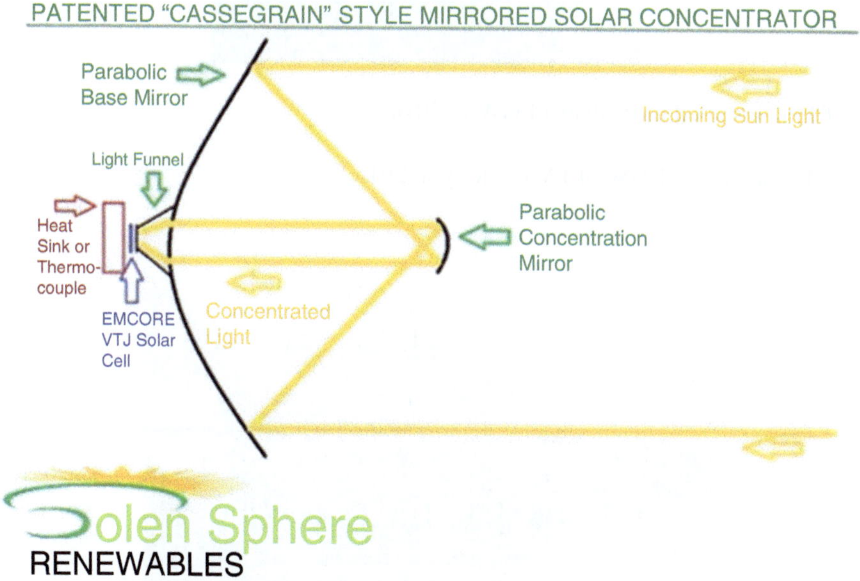

Fig. 1.2 Various advances in PV technology while Fig. 1.3 shows recent progress in PV utilization

Fig. 1.3 Figure 3 shows recent progress in PV utilization

The University of Bahrain and BAPCO has installed 5 MW of PV in Bahrain.

China installed more than 14 GW in 2014

UK reaches 8.31 GW of PV capacity in 2015 [3].

US has installed 13.48 GW in 2015.

Germany has installed a total of 39.553 GW by the end of 2015.

China has total installation in September 2015 s 37.95 GW.

Nineteen rooftop PV systems with a capacity of over 30 MW will be installed at the port of Jebel Ali, Dubai, UAE. The PV systems will cover one-third of the port's electricity needs and will be installed on parking sheds and in the surrounding parking areas. It is ready by 2016 [4].

The Prime Minister Narendra Modi has set up a target of 100 GW by 2022.

More than 1,463,867 rooftop PV systems, representing 4.47 GW of installed PV capacity, have been installed in Australia up to last September 2015.

The Netherlands has reached 1.32 GW of PV capacity in 2015.

Switzerland has reached1.35 GW in December 2015 [5].

Over the last 25 years, the yearly production growth of PV has been more than 30 % per annum; see Fig. 1.4.

Figure 1.5 shows that the bulk of PV production in 2015 was of multi-crystalline silicon cells, of which multi-Si = 53 GW and mono Si = 9.5 GW, while thin films were approximately 6 % with 4.2 GW. Production per country as percentage are the following: Taiwan 24 %, China 22 %, USA 16 %, Japan 16 %, Korea 2 % and EU 10 %. [7]

In terms of cost, Fig. 1.6 shows the progressive decrease in PV cost from 1977 to date. The pay back of any PV system (1 kW–1 MW) is 1.7 year.

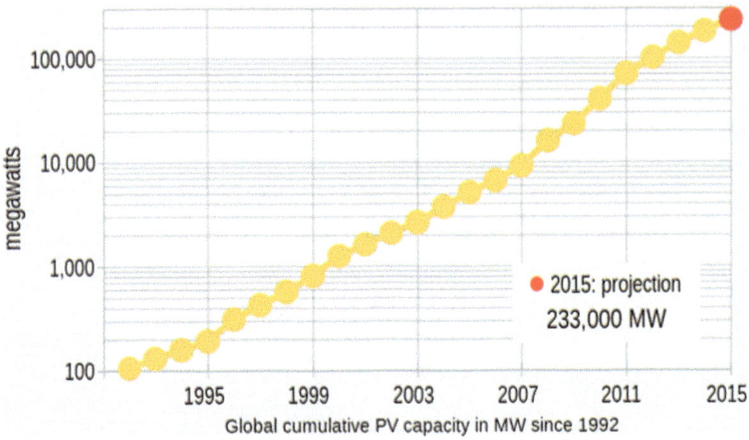

Fig. 1.4 PV growth from 1990 to 2015 [6]

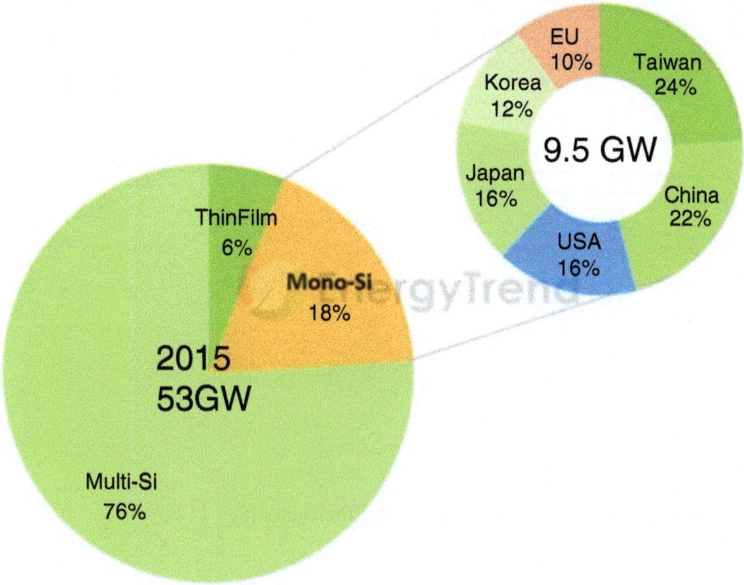

Fig. 1.5

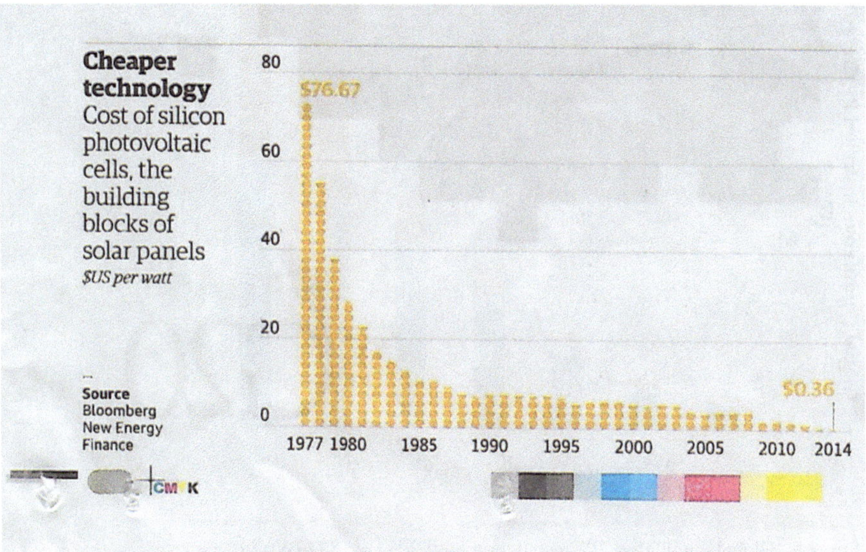

Fig. 1.6

Fraunhofer Institute in their report: Photovoltaic Report 17 November 2015 Stated of PV system (10 kW–100 kW) in 2015 is US$1.5 per Watt. Generally the cost can be divided into cost of BOS including inverter is 52 % and module cost is 48 %.

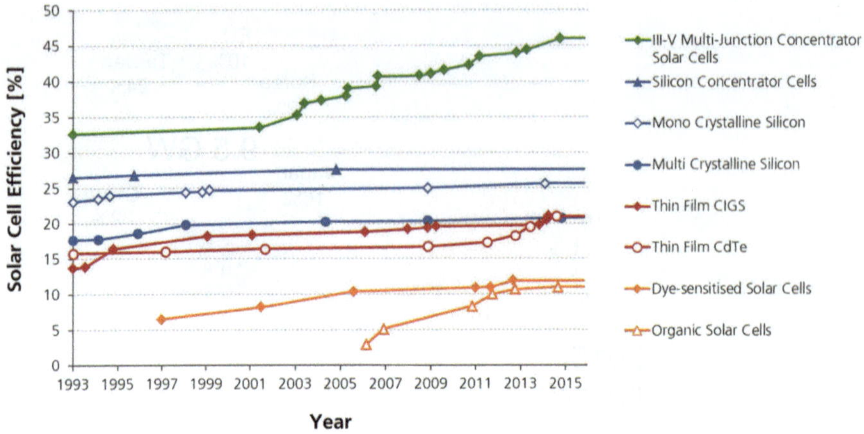

Data: Solar Cell Efficiency Tables (Versions 1-46), Progress in Photovoltaics: Research and Applications, 1993-2015. Graph: Simon Philipps, Fraunhofer ISE 2015

Fig. 1.7 Development of Laboratory Solar Cell Efficiencies [8]

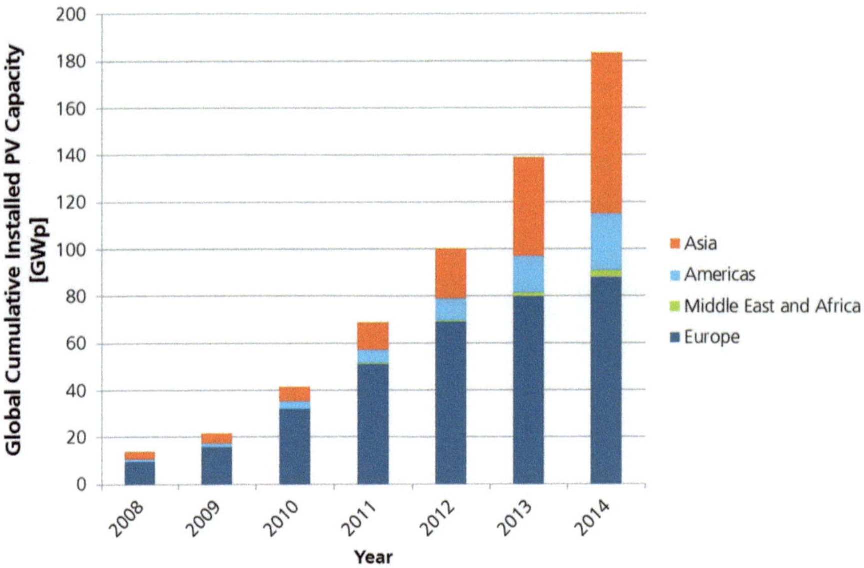

Fig. 1.8 Global cumulative PV installation until 2014 = 183 GWp [9]

References

1. Global Energy World (2014), Accessed 19 Nov 2014
2. World of Renewable (2012), Accessed 12 Sept 2012
3. Photon Newsletter (2015), Accessed 27 Nov 2015
4. PHOTON Newsletter (2015), Accessed18 Sept 2015
5. PHOTON Newsletter (2016), Accessed 13 Jan 2016
6. https://en.wikipedia.org/wiki/Photovoltaics
7. Energy Trend (2015), Accessed Oct 2015
8. Solar cell efficiency tables (Versions 1-46), Progress in photovoltaics: Research and applications, 1993–2015. Graph: Simon Philipps, Fraunhofer ISE 2015
9. IHS. Graph: PSE AG 2015

Chapter 2
The Bell Telephone Laboratories Discovery: Ushering in Our Modern Age of Solar Photovoltaics

Lawrence L. Kazmerski

2.1 Introduction: The Setting

> New York Times, April 26, 1954: *MURRAY HILL, N.J.—A solar battery, the first of its kind, which converts useful amounts of the sun's radiation directly and efficiently into electricity, has been constructed here by Bell Telephone Laboratories . . . It may mark the beginning of a new era, leading eventually to the realization of one of mankind's most cherished dreams—the harnessing of the almost limitless energy of the sun for uses of civilization.*

Charles Dickens [1] had it half correct, "it was the best of times" for science in this 1950s decade of discovery. The VCR, the bar code, radial tire, music synthesizer, transistor radio, Teflon, the pill, optical fiber, the laser, computer hard disk, black box flight recorder, pacemaker, Fortran, the modem, Velcro, and the integrated circuit all appeared during these ten years. DNA was discovered, as well as the neutrino and the polio vaccine. *(Aside: Some might consider the "1950s" also tilted toward the "worst of times" with introductions of the credit card, McDonalds, the Barbie doll, the TV remote control, Mr. Potato Head, the hula hoop, and the snowmobile!)* Bell Telephone Laboratories was nurturing its world-class science following its game-changing invention of the transistor in the late 1940s [2]. Their success was built on the perfect storm for creativity and innovation—excellence in staff (experienced, accomplished, and supportive technical managers who embraced the success of their staff as their success as well, best-of-class scientists, engineers, technicians), outstanding facilities, and empowered and recognized research teams. "It was the best of times..."

L.L. Kazmerski (✉)
National Renewable Energy Laboratory (Emeritus), Golden, Colorado 80401, USA

Renewable and Sustainable Energy Institute, University of Colorado, Boulder, CO 80309, USA
e-mail: solarpvkaz@gmail.com

A. Sayigh (ed.), *Photovoltaics for Sustainable Electricity and Buildings*,
DOI 10.1007/978-3-319-39280-6_2

In April 1954, the announcement came at the NY Headquarters of the National Academy. With the media spotlight focused brightly on this new invention from Bell Telephone Laboratories, the crystalline-Si solar cell ("the solar battery") was rolled out to the public, with a sunlight-to-electricity conversion efficiency of about 6 % [3, 4]. The New York Times (see above) marked it as the "beginning of a new area" with "limitless energy." US News & World Report [5] envisioned that these silicon strips "may provide more power than all the world's coal, oil and uranium." We are now finally on our way to fulfilling this prophesy—to create this clean energy future.

2.2 Why at Bell? Why Then?

This discovery was not serendipitous. Nor did Daryl Chapin, Gerald Pearson, and Calvin Fuller (iconic photo in Fig. 2.1) plan on founding a solar industry. Bell was growing its revolutionary semiconductor development built on the solid-state rectifier and the transistor. And, Bell Labs had a problem—that of powering telephone repeaters in remote areas having relatively harsh climate conditions. In the US southwest, batteries degraded rapidly and were costly to replace. The Labs challenged its scientists and engineers—and teams were formed to investigate several potential solutions (steam engines, wind, thermoelectrics, among others) [3]. *The Lab was committed; scientists were empowered.*

Fig. 2.1 Iconic photo of the Bell Inventors, Gerald Pearson, Daryl Chapin, and Calvin Fuller, testing their "solar battery" in 1954 (with permission of Bell Laboratories)

2.3 The Bell Solar Battery Story in Brief

Though Bell Labs had previous solar cell R&D and IP (including work on Si), this was a tipping point. The response started in 1952 with Chapin, who was investigating selenium solar cells—a PV material which has its origins in the late 1800s. But Chapin's work was frustrating, only able to convert <1 % of the light to useful electricity. A colleague pointed him to Gerald Pearson, who was working on semiconductor Si with some interesting initial results. Calvin Fuller, whose work on diffusion had laid the foundation for solid-state rectifiers, guided Pearson to use a hot lithium bath to dip his p-type (Ga-doped) Si to form a shallow junction. The dream team was formed: the *physicist* (Pearson), the *engineer* (Chapin), and the *chemist* (Fuller). Chapin took the Pearson-Fuller rectifier—and confirmed that lithium-gallium rectifies performed superior as a solar cell to anything that had been reported previously—in the 2 %-efficiency range. Pearson directed the group to switch to Si—and avoid "wasting another moment" on Se. Chapin's notebook showed his calculation that 23 % conversion efficiency was attainable—and that the group established a "6 % efficiency goal" to meet the Bell power requirements (Timeline in Fig. 2.7). *The race for a viable solar power source was on.*

The optimism of the initial success soon sank. Problems emerged with contacts and the unstable lithium-based junction. Moreover, a junction that Chapin diagnosed to be too deep below the surface because the Li-produced junction moved into the Si even at room temperatures: "It appears necessary to make our p–n junction very near to the surface so that nearly all the photons are effective" (Chapin Lab-Notebook entry). Enter the chemist. *First*, Fuller introduced arsenic into the silicon crystal during growth to make it n-type. *Then* to stabilize and control junction depth, he applied the diffusion processes he had pioneered with the transistor, using boron gas to form the shallow and uniform p-type layer. Contacting these more higher-doped and stable layers was easier. Fuller's process improved the cell performance, and the team reached the 6 % efficiency goal in January 1954 [4]. John Ditzenberger, Fuller's technician at the time, quotes Pearson declaring "Silicon is the material and diffusion is the process!" With this baseline, the Bell team began assembling these adolescent Si cells into small modules for demonstration of this perfected technology. It was almost ready to be shown to the public—and *Bell Labs was really good at this.*

2.4 Bell Was Not Alone: The RCA Rivalry

But there was competition from nearby Princeton, New Jersey, and RCA Laboratories. The American *Atoms for Peace* program introduced to the United Nations in December 1953 by President Dwight Eisenhower promoted using nuclear for energy instead of bombs. Paul Rappaport with Joseph J. Loferski developed "Betavoltaic Cells," p–n junction devices designed to provide power from nuclear waste radiation [6, 7]. But in testing, this "atomic battery" had to be enclosed in a dark box because the light from the windows overwhelmed the effect from their nuclear source! RCA rolled out their invention with a demonstration by David Sarnoff—using this new atomic battery to power a telegraph to tap out "Atoms for Peace" in code in January 1954. *The pressure was on.*

2.5 Bell Scientists Succeed

With the baseline processing established, the Bell team began assembling these adolescent Si cells into small modules for demonstration of this perfected technology. In late 1953, Pearson brought in another bright young postdoc who had been hired out of MIT to work on Bell's transistor project a year before by William Shockley—Morton B. Prince. Prince (Fig. 2.5) became a member of staff and was given the responsibility to oversee the module work (Fig. 2.4) and to develop projects to show the effectiveness of these new solar batteries. He developed solar battery-powered radio transmitters, radio receivers, toys, and other gadgets. Mort Prince also readied the solar battery and its physics for scrutiny by the technical community. His seminal paper in 1955 on the performance expectations of the cell remains one of the most cited early references in PV [8].

But first, the solar battery was ready to be unveiled to the public—and Bell Labs was really good at this. The 1954 press event in New York was a media success. On April 25, 1954, Bell executives presented the "Bell Solar Battery" to the public with a display of cells using only light to operate a 21-inch Ferris wheel (assembled from an Erector Set, Fig. 2.2). A reporter from Popular Electronics [9] noted with the Bell solar cells linked together could "deliver power at the rate of 50 W per square yard

Fig. 2.2 Bell Ferris wheel (constructed from an Erector Set—used to demonstrate the solar battery (*lower left of wheel*) at the National Academy news conference in April 1954

while the RCA atomic cell delivered only a millionth watt over the same area, a difference of 50 million." *Game and match—Bell Laboratories!*

2.6 Operation, Outreach, Outer Space

A sometimes-overlooked aspect of the solar battery evolvement is how quickly Bell Labs was able to take its lab bench discovery to use in the field. In October 1955, the first module was commissioned in Americus, Georgia (Fig. 2.3) [3]—*less than 2 years after cell proof of concept*. The modules were about 35 W, assembled from submodule components that sealed the cells in oil for optical coupling and protection (perhaps a valid demonstration of oil sequestration!). The Americus module was tested at National Renewable Energy Laboratory (NREL) in 2004 and still had an output of 16 W. The repeater-power project was a technical success (combining the solar battery with the lower power requirements of transistor electronics), but with the panels costing about $1000/W, the economics could not sustain further use. Interestingly, the Americus module had disappeared from Bell archives—but had

Fig. 2.3 Americus, Georgia module used to power telephone repeater, being lab assembled, showing submodule of cells encased in oil (*top*) and deployed in Southwest US location (*bottom*)

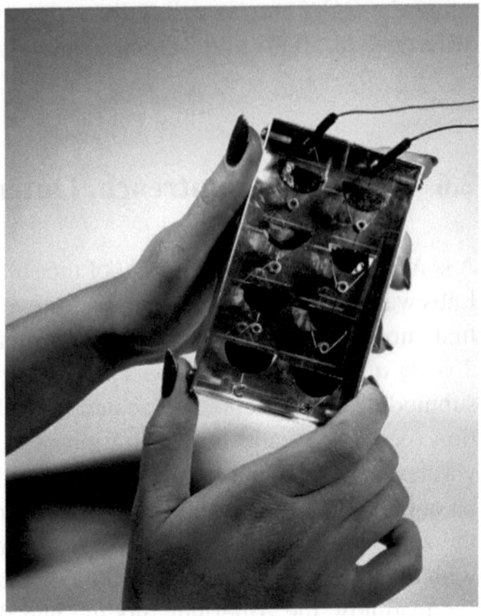

Fig. 2.4 Module designed and fabricated by M. Prince in 1954. (For the record, those are not his hands!)

been put to clever use by Chapin in his New Jersey home garden to power an electric fence to ward off vegetable-loving raccoons [Fig. 2.6]!

Bell Laboratories was said to have fostered more publicity about the solar battery than for the transistor. It helped fuel a growth in interest at this very time in solar energy. In the fall of 1955, the first large international [10] brought together scientists, inventors, and technologists from around the world for a coordinated, two-part meeting. The "Conference on Solar Energy—The Scientific Basis" was held in Tucson, Arizona, October 31 to November 1, 1955; and the Solar Energy Conference and Exhibit was held in Phoenix—giving birth to the America Solar Energy Society (ASES) and the International Solar Energy Society (ISES)—and introduced the potential for solar power on earth. Bell played a big part at this meeting. Mort Prince was handed the responsibility to present the physics and technology behind the Bell cell in a paper "Silicon p–n Junction Solar Energy Converter" in Tucson. Pearson, Chapin, and Prince headlined the photovoltaics portions of the meetings and the exhibit.

In the mid- to late 1950s, PV entered the business sector. Hoffmann Electronics coaxed Prince away from Bell to become its director of research—and to initiate work and commercialize the Bell solar battery. An impressive array of PV pioneers (Martin Wolf, Eugene Ralph, Peter Iles) provided the drive for solar power used in space—and specifically for the first PV-powered satellite, Vanguard, in March 1958 [R]. The Bell Discovery was foundational for our PV technology. *The Bell Trio were the right scientists at the right time, at the right place. We owe them and Bell so much.* (Figs. 2.4, 2.5, 2.6 and 2.7)

Fig. 2.5 Dr. Morton B. Prince, Bell Labs member of staff in 1954

Fig. 2.6 Dr. Chapin and his PV-powered raccoon-repelling electric fence!

SOLAR BATTERY INVENTOR — Daryl M. Chapin with duplicate of the original solar battery which powered the experimental telephone line in Georgia. This solar battery in his garden charges a battery which powers an electric fence to keep prowling animals out of his garden. (Staff photo by Marko.)

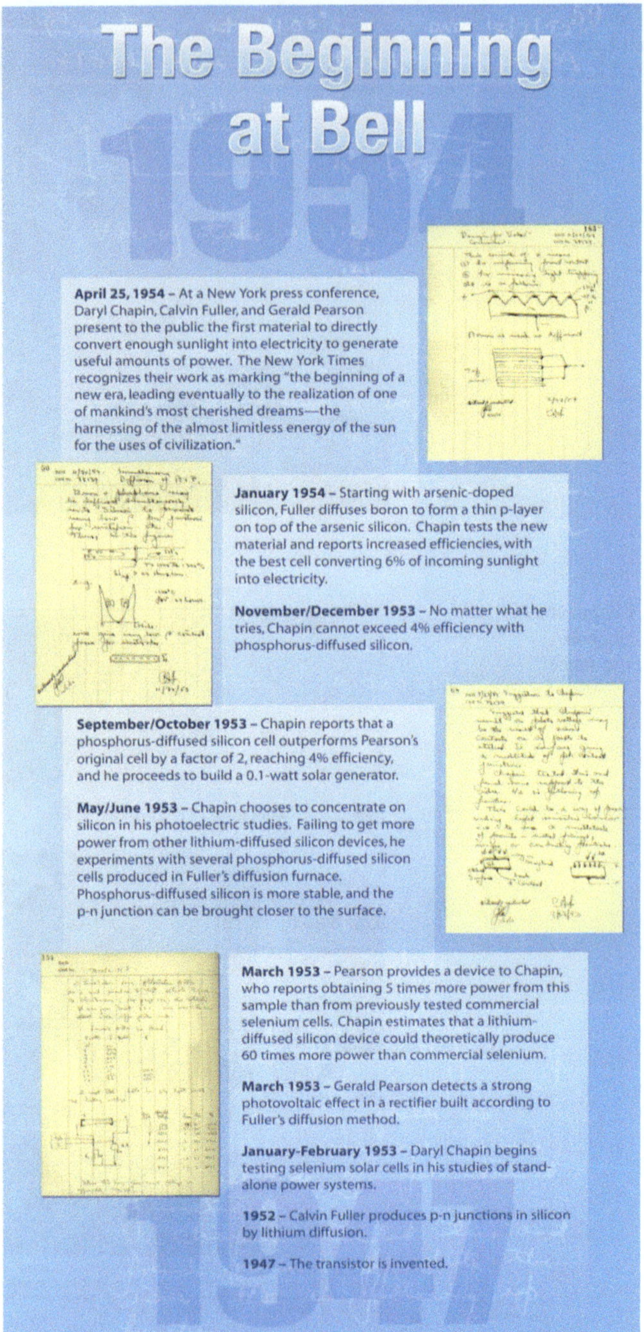

Fig. 2.7 Bell Telephone Laboratories discovery timeline (from 50th anniversary exhibit by NREL), with pertinent lab notebook entries

Acknowledgments The author gratefully acknowledges Dr. Morton B. Prince for his generous guidance, memories, artifacts, and inputs—and who at this time is the last remaining member of the historic team that gave us the modern solar cell. The author also acknowledges the many interviews from the pioneers in PV whose videos formed the basis for the Bell Labs 50th anniversary exhibit that was unveiled in the WCPEC in Osaka, 2004. This paper is partially based upon the research that went into that exhibit and the preparation of Ref. [3] (with colleagues John Perlin and Susan Moon). The author also wants to especially remember the help of Audrey Swenson who provided many of the inside stories of her father (Daryl Chapin)—and also the late John Ditzenberger (RIP 2004), whose down-to-earth descriptions of Bell working environment and the famous trio give insight to the greatness of Bell Labs and the humanity and excellence of Daryl Chapin, Calvin Fuller, and Gerald Pearson.

References

1. C. Dickens, *Tale of two cities* (Chapman and Hall, London, 1859)
2. www.nobelprize.org/educational/physics/transistor/history
3. John Perlin, Lawrence L. Kazmerski, Susan Moon, Good as gold: The silicon solar cell turns 50. Solar Today, 24–27 (Jan/Feb 2004)
4. D.M. Chapin, C.S. Fuller, G.L. Pearson, A new p- n junction photocell for converting solar radiation into electrical power. J. Appl. Phys. **25**, 676 (1954)
5. U.S. News and World Report, (April 25, 1954)
6. P. Rappaport, The electron-voltaic effect in p-n junctions induced by beta particle bombardment, Phys. Rev. **93**, 246 (1953)
7. P. Rappaport, J.J. Loferski, E.G. Linder, The electron-voltaic effect in germanium and silicon p-n junctions. RCA Rev **17**, 100 (1956)
8. M.B. Prince, Silicon solar energy converters. J.Appl. Phys. **26**, 534–540 (1955)
9. Popular Electronics, Sun rays used in new invention. (October 1954)
10. K.W. Boer, (ed.), The fifty-year history of the International Solar Energy Society. (ISES, Germany; 2005) ISBN 10: 0895533251 ISBN 13: 9780895533258

Chapter 3
Effective Solar Resource Methodologies for Sustainable PV Applications

David S. Renné

3.1 Introduction

Through focused R&D efforts, favorable policies, and access to private capital financing, solar photovoltaic (PV) technologies have experienced phenomenal growth in the marketplace during the past three decades. During this time, the global installed capacity of PV has doubled every 3–4 years, so that at the time of this writing, global capacity is approaching 300 GW [1]. This growth is attributed to the positive feedback cycle of improved efficiencies and reliability resulting from public and private R&D investments, which brings down the commercial costs and promotes positive policy making to further expand markets and encourage new investments. One major factor in stimulating the growth of this key carbon-free technology has been the constant improvement in understanding of the solar resource available to these technologies. Solar resource availability and its characteristics are among the most important factors to be considered as further investments are contemplated. During the past three decades, as R&D in improving cell and panel efficiencies and reliability has paid dividends, so has the concurrent R&D in improving our understanding of the solar "fuel" available to these technologies. This understanding not only provides guidance on optimal sites for which to install systems but also on how to optimize the design and operation of systems at any site.

Because the solar resource available to photovoltaic technologies, whether they are stand-alone systems or integrated into buildings, is highly variable, both in time and in location, this chapter focuses on how we can obtain the best possible knowledge about this variability so that photovoltaic systems can be designed and deployed optimally regardless of the load they are being used to meet. Notwithstanding the nocturnal and seasonal planetary cycles that impact the solar resource

D.S. Renné (✉)
Dave Renné Renewables, LLC, 2385 Panorama Avenue, Boulder, CO, USA
e-mail: drenne@mac.com

A. Sayigh (ed.), *Photovoltaics for Sustainable Electricity and Buildings*,
DOI 10.1007/978-3-319-39280-6_3

reaching the top of the atmosphere, weather patterns, landform characteristics, and topography all play key roles in affecting the amount and short-term variability of the solar resource.

PV technologies can be characterized as **variable renewable energy** (VRE) resources. Solar resource variability influences the **net load** (the difference between the utility load and the concurrent energy supplied by PV to meet that load). VRE, therefore, influences the utility's approach to meet the system load in ways that may be remarkably different from the traditional way that utilities without VRE have addressed changing load profiles.

Despite the importance of a clear understanding of resource characteristics for ever-increasing PV penetrations into a grid, actual long-term reliable on-site ground-based measurements are often lacking at most proposed sites. Thus, various tools and techniques need to be brought into play in order to best characterize the long-term resource so that investment risk in the project is minimized.

This chapter introduces current research and best practices addressing the influence of the solar resource on effective PV applications. Some of the information provided in this chapter has come out of an international collaborative effort working under the International Energy Agency's Solar Heating and Cooling Implementing Agreement. The collaboration most recently has been working on a task titled "Solar Resource Assessment and Forecasting." Outcomes from this collaboration can also be found at http://task46.iea-shc.org.

3.2 The Components of Solar Resources

The solar radiation (or irradiance) reaching the top of the earth's atmosphere is referred to as the **total sky irradiance** or extraterrestrial radiation and is currently estimated at 1366 W-m^{-2}. The energy emitted by the sun is virtually invariant (approx. 0.1 %), but because of the eccentricity of the earth's orbit, extraterrestrial radiation varies by about 6.7 % throughout the course of a year [2].

The sun emits radiation across a broad spectrum of wavelengths, as shown in Fig. 3.1, which gives the spectral distribution of radiant energy reaching the top of the earth's atmosphere (taken from [2]). The portion of the solar spectrum that is visible to the human eye is generally in the range of 380–780 nm, which is also the range in which the radiance emitted by the sun is at a maximum (the absolute maximum is at around 500 nm). This portion of the spectrum is called the **visible spectrum**. Radiation in wavelengths shorter than 380 nm is known as **ultraviolet** radiation, and radiation occurring in wavelengths longer than around 780 nm is called **infrared** radiation. The majority of solar radiation reaching the top of the earth's atmosphere falls in the spectral range of 300 nm to 3000 nm.

The gases that make up the atmosphere can both absorb and scatter the incoming solar radiation at a number of discreet wavelengths. Thus, the spectral distribution of the radiant flux reaching the earth's surface can look quite different from the distribution shown in Fig. 3.1. For example, most of the ultraviolet region is

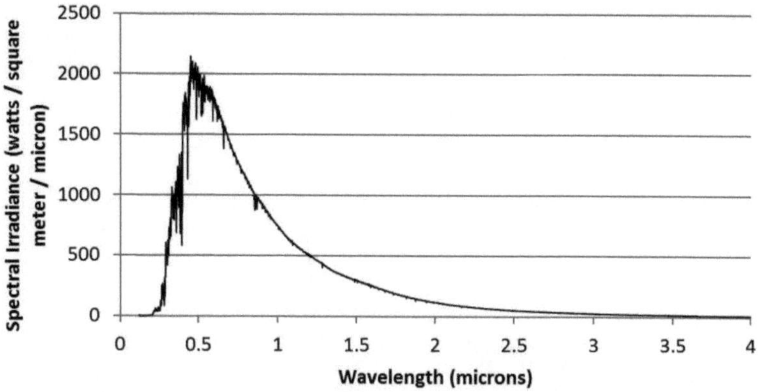

Fig. 3.1 The spectral distribution of solar radiation reaching the top of the earth's atmosphere (from [2])

absorbed and scattered by atmospheric molecules (especially ozone), so that the amount of ultraviolet radiation at the earth's surface is quite low. The reason the sky appears blue to the human eye is due to the preferential scattering of air molecules in the blue wavelength region of the visible spectrum. Atmospheric water vapor absorbs solar energy in a number of different wavelengths, especially in the infrared regions. Other trace gases, such as CO_2 and methane, can also absorb solar radiation in a variety of spectral ranges, causing the temperature of the atmosphere to increase. Nevertheless, at the earth's surface, the predominant wavelength range for the solar resource is 0.4–0.7 μm (400–700 nm), which also happens to be the predominant wavelength region to which the majority of PV systems respond.

The various **components** of the solar resource are explained in Fig. 3.2. The definitions apply primarily to the amount of radiation reaching the earth's surface in the visible spectrum (400–700 nm).

The amount of solar radiation from the sun reaching a flat surface at the earth's surface oriented normally to the sun's position throughout the day is known as **direct normal** irradiance, or **DNI**. Portions of the visible radiation reaching the top of the atmosphere can be scattered back to space or reflected toward the earth's surface by the presence of cloud droplets and aerosols. Thus the total radiation reaching the earth's surface will be a combination of the **DNI**, the amount re-emitted by air molecules, as well as the radiation scattered back to the surface due to cloud droplets and aerosols. This portion of the sky radiation is known as **diffuse** radiation. Measuring this component on a flat surface oriented horizontally gives the **diffuse horizontal irradiance**, or **DHI**. If we were to measure the total radiation falling on a flat surface horizontal to the earth's surface, we would be measuring both the solar radiation coming directly from the sun (assuming it is not obscured by a cloud) and the diffuse radiation from the clouds and sky. This, then, is the **global horizontal irradiance**, or **GHI**.

GHI represents the total energy available from the sun and the sky on a horizontal surface and is usually the most important factor for understanding the

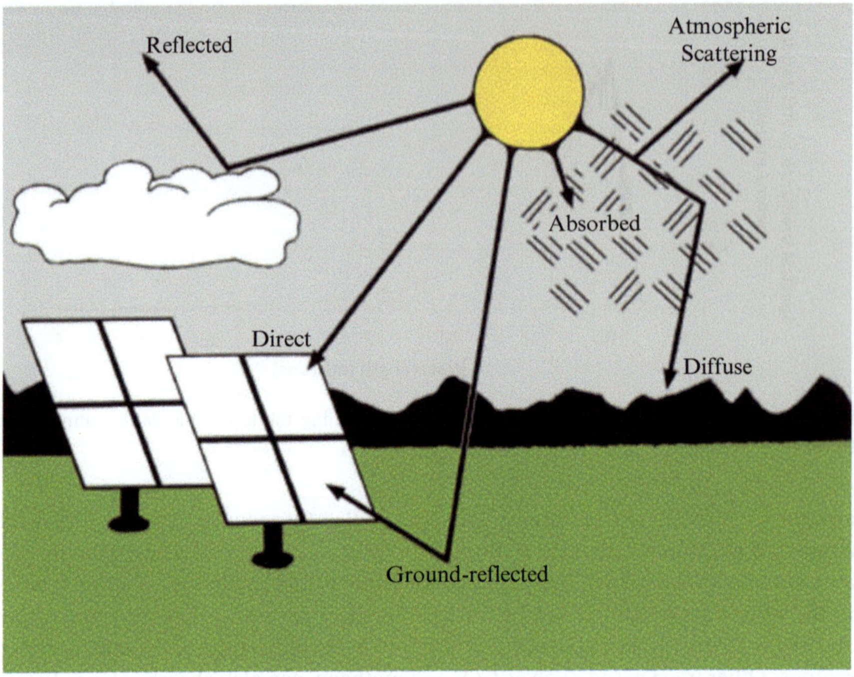

Fig. 3.2 The components of the solar resource (source: NREL)

resource available to flat plate collectors such as those used in the majority of PV panels. DNI, on the other hand, is the most important component of solar radiation for understanding the energy that can be produced by concentrating collectors, such as curved mirrors used in troughs, or concentrating PV technologies. Mathematically, the relationship between GHI and the other two components, DHI and DNI, is given as

$$\mathrm{GHI} = \mathrm{DNI} * \mathrm{Cos}(Z) + \mathrm{DHI} \tag{3.1}$$

where Z is the **solar zenith angle**, or the angle between the sun's position at any given time and the zenith, or a vertical line perpendicular to the earth's surface.

In most practical applications, flat plate collectors are generally oriented more directly toward the sun rather than on a horizontal plane. This is because a collector oriented perpendicular to the sun will receive a higher global irradiance flux than on horizontal surfaces (except, of course, when the sun is directly overhead, at the zenith). Often, fixed flat plate collectors are oriented at an angle toward the south (in the northern hemisphere) equal to the latitude where the collector is installed. This is known as **latitude tilt**. In some cases flat plate collectors might even be mounted on one-axis or two-axis trackers to allow the collector to be oriented normal to the sun throughout the day and throughout the seasons. The conversion of GHI to the irradiance received on a tilted surface is called **global tilt irradiance**

(GTI). As noted in Sengupta et al. [3], a number of models are available that convert GHI or a combination of GHI and DNI information into GTI, such as Temps [4], Hay [5], Kluchar [6], Liu and Jordan [7], Reindl [8], and Gueymard and Myers [9]. A simple method to convert monthly and annual average GHI to GTI for one- or two-axis trackers as part of the products developed from the US National Solar Radiation Database (NSRDB), 1961–1990, is described in Marion and Wilcox [10]. This reference also provides a volume of data values of these conversions for the 239 stations making up the 1961–1990 NSRDB.

3.3 Understanding the Solar Resource for PV Operations

For all solar technology applications, there are two primary means of obtaining solar resource data at a given location: (1) direct measurement of the solar resource using qualified instrumentation and accepted measurement practices and (2) modeling or estimating the solar resource using data sources such as imagery of cloud characteristics observed from weather satellites, ground-based cloud cover observations, numerical weather prediction models, or sunshine duration records. This section provides an overview of these two approaches.

3.3.1 Solar Resource Measurement Techniques

Although efforts to measure the solar resource quantitatively have been around for nearly two centuries (such as recording the number of hours of direct sunshine or the use of bimetallic strip recorders), at the beginning of the twentieth century, methods that convert the incoming solar radiation to an electrical output were developed, and today this is the most common approach for developing precise measurements of the solar resource. Instruments using this approach fall into two broad categories: those using **thermopile**-type detectors and those using **silicon-photodiode**-type detectors. A thermopile detector works on the principle of the **thermoelectric effect**, where a voltage is generated from the temperature difference between two dissimilar metals. Today's precision solar monitoring devices make use of this effect by deploying two different metals under a glass dome (or double-glass dome) and monitoring the voltage output of the incoming solar radiation, either from the total sun and sky (**pyranometers**) or directly from the solar disk (**pyrheliometers**).

Silicon photodiodes are made from crystalline silicon that has been transformed into a semiconductor, similar to the principles of a crystalline silicon PV cell. Instruments using this technology have the advantage of much lower cost and much faster response time than a thermopile-based instrument; their main disadvantage is their limited spectral response characteristics (they do not respond at all

(a) (b) (c)

Fig. 3.3 Examples of commercially available solar monitoring devices: (**a**) Eppley pyranometer, (**b**) Eppley pyrheliometer mounted on a tracker, and (**c**) LiCor silicon photodiode sensor (source: NREL Image Gallery # 15537, 15554, and 15483)

to wavelengths above 1100 nm). This disadvantage results in a slightly higher uncertainty than instruments using thermopile detectors.

Examples of a thermopile-based pyranometer and pyrheliometer currently on the market are shown in Fig. 3.3a, b; an example of a silicon photodiode pyranometer currently on the market is shown in Fig. 3.3c.

The use of thermopile radiometers became common beginning in the mid-1920s due to the innovative work of companies such as Kipp & Zonen and, later, Eppley Laboratories and EKO Instruments, which continue to develop and refine improved thermopile instruments for a variety of scientific purposes. Other companies such as Hukseflux and Yankee Environmental Systems have entered the market more recently with thermopile-based products. Instruments based on silicon photodiode technology are also very common, due to their low cost.

The choice of instruments used in a solar resource measurement campaign depends on the uncertainty goals established for the project. Currently, the World Meteorological Organization (WMO) and International Organization for Standardization (ISO) 9060 categorize pyranometers and pyrheliometers according to their accuracy and uncertainty. As shown in Sect. 3.2 of Sengupta et al. [3], the WMO categorizes pyrheliometers into two classes, high quality and good quality, and pyranometers into three classes, high quality, good quality, and moderate quality. The ISO 9060 specifications for both pyrheliometers and pyranometers fall into three classes: secondary standard, first class, and second class. ISO secondary standard (or WMO first class) typically has the shortest response time, the highest resolution and stability, and the lowest uncertainty of the instruments available on the market. Research-grade measurement programs typically use secondary standard (WMO high quality) instruments. Sites being monitored for solar resource potential often deploy first-class (WMO good quality) sensors although, as projects become larger and larger and bank financing requirements become more stringent, some developers are now opting for secondary standard equipment. Secondary standard instruments are delivered by the manufacturer with calibrations traceable to the World Radiometric Reference (WRR), which is the globally accepted standard established every 5 years at the World Radiation Center in Davos,

Switzerland. Thus, secondary standard instruments are typically more expensive to purchase than first-class instruments and should be recalibrated in the field at least annually with reference instruments traceable to the WRR.

In order to obtain the most accurate complement of DNI, GHI, and DHI at a site, individual thermopile-based instruments must be used for each component, i.e., a pyrheliometer on a tracking device for DNI and two pyranometers (one an unshaded instrument measuring GHI and the other with a shading device that blocks the direct beam so that it measures only DHI). Because of the expense of such a configuration, there have recently been efforts to utilize a simpler and less costly but still reliable instrument, known as a **rotating shadowband radiometer**, or **RSR.**[1] The RSR basically consists of a silicon photodiode sensor mounted on top of a box that contains a motor and a shading band (see Fig. 3.4a, b). The motor is designed to cause the shadow band to periodically pass over the field of view of the sensor. Because of the fast response time of the sensor, as soon as the shading band completely shades the sensor, the sensor is measuring only the DHI. At all other times, the sensor is measuring the GHI. Based on these two measurements, the DNI can be calculated by solving Eq. (3.1) for DNI:

$$\mathrm{DNI} = (\mathrm{GHI} - \mathrm{DHI})/\mathrm{Cos}Z \tag{3.2}$$

There has been extensive research undertaken to evaluate the uncertainties of thermopile-based pyrheliometers and pyranometers operated in the field, as well

(a) (b)

Fig. 3.4 Images of a rotating shadowband radiometer (RSR): (**a**) close-up of the shading arm passing over a silicon-diode sensor and (**b**) RSR installed in the field (source: NREL Image Gallery) #15484 (**a**) and #16994 (**b**)

[1] In some cases these instruments are referred to as rotating shadowband pyranometers (RSPs) or rotating shadowband irradiometers (RSIs).

as the uncertainties in operational RSRs. For example, extensive comparative studies have been conducted at the National Renewable Energy Laboratory's Solar Radiation Research Laboratory in Golden, Colorado, and these studies have been published in Wilcox and Myers [11], with a recent update published by Habte, Wilcox, and Stoffel [12]. A summary of uncertainty analyses for pyrheliometers and pyranometers is also provided in Sengupta et al. [3]. This report shows that pyrheliometers operated in the field following proper maintenance will have a sub-hourly uncertainty at the 95 % confidence level of ±2 %, while pyranometers will have an uncertainty of ±3 % for solar zenith angles between 30° and 60°, and up to ±7–10 % for solar zenith angles greater than 60^0. The uncertainty of an RSR DNI measurement at the 95 % confidence level is undergoing considerable research; in Vignola et al. [2], the uncertainty is reported at ±3.2 %. Geuder et al. [13], which is also reported in Sengupta et al. [3], reported an uncertainty of about ±3 % based on a study of 39 RSRs tested at Plataforma Solar de Almería, Spain. Generally the uncertainties of an RSR are expected to be somewhat higher than with the thermopile instruments, although from the perspective of field operations these slightly higher uncertainties may be offset by their significantly lower costs.

Other methods to develop a low-cost approach for measuring DNI are currently being tested and commercialized. One such instrument is the SPN-1, produced by Delta-T Devices (Fig. 3.5), which is capable of providing DNI, DHI, and GHI without the use of any moving parts. As described in Sengupta et al. [3], this instrument consists of an array of seven thermopile detectors distributed under a glass dome in a hexagonal pattern. The sensors are located underneath a series of diffusers and a shadow mask, which is designed in such a way that throughout the day there will always be one or more detectors that are fully shaded from the sun and exposed to approximately half of the diffuse solar radiation (under overcast skies). Also, one or more detectors are exposed to the full solar beam for all solar

Fig. 3.5 The SPN-1 pyranometer produced by Delta-T Devices (source: Delta-T, UK)

positions. The minimum and maximum readings of the seven detectors are then used to calculate GHI and DHI. Although at present the uncertainties of the outputs of this instrument are somewhat higher than the thermopile instruments or even the RSRs, based on testing at NREL and at PSA, work continues to improve the accuracy of this instrument. This instrument represents a promising low-cost approach to provide measurements of all three of the solar components from a single device with no moving parts.

3.3.2 Estimating the Solar Resource from Weather Satellite Observations

In the early 1980s, when **geostationary weather satellites** (weather observation satellites that orbit at the same speed as the earth's rotation, so that they are always positioned 40,000 km above the equator at a fixed longitude) came into common use, studies to use the visible channel of the satellite to develop an estimate of the solar insolation at the earth's surface were first undertaken. Because of their fixed location above the earth's surface, these satellites have the advantage of being able to monitor cloud conditions almost continuously (every 15 min with modern-day satellites), at a resolution of about 1 km in the visible channel at the subsatellite point. Because of their position above the earth, however, their effective "view" is between 60° N and 60° S. About five satellites are needed around the world to obtain a complete view of the earth's surface within this latitude band.

Weather services also make use of **polar orbiting** satellites, which orbit the earth at low altitude (approx. 400 km) from pole to pole. Polar orbiting satellites have the advantage of being able to observe the earth's clouds and surface at much higher spatial resolution than geostationary satellites, due to their much lower orbit. However, because the earth is rotating under these satellites, a satellite generally passes over the same point on earth's surface at most only twice a day, making the time resolution of observing cloud patterns much coarser than with geostationary satellites.

Figure 3.6 shows the current location of geostationary satellites, which are operated by several countries, around the world. The figure also shows the typical coverage that is obtained from a polar orbiting satellite. The majority of the methods developed for solar resource assessment make use of geostationary satellite imagery, although there have been some studies that include polar orbiting satellites to obtain complete coverage (e.g., see the NASA Surface Meteorology and Solar Energy Dataset at https://eosweb.larc.nasa.gov/sse/).

The approaches developed to produce solar resource estimates from satellite observations can be categorized in three basic ways: **empirical** methods, where the strength of the visible signal at the satellite is correlated with various high-quality ground-based measurement stations; **semiempirical** methods, where the strength of the visible signal is compared with the ground albedo (as determined by monitoring the signals at the satellite on clear days) to produce a "cloud index" from which a

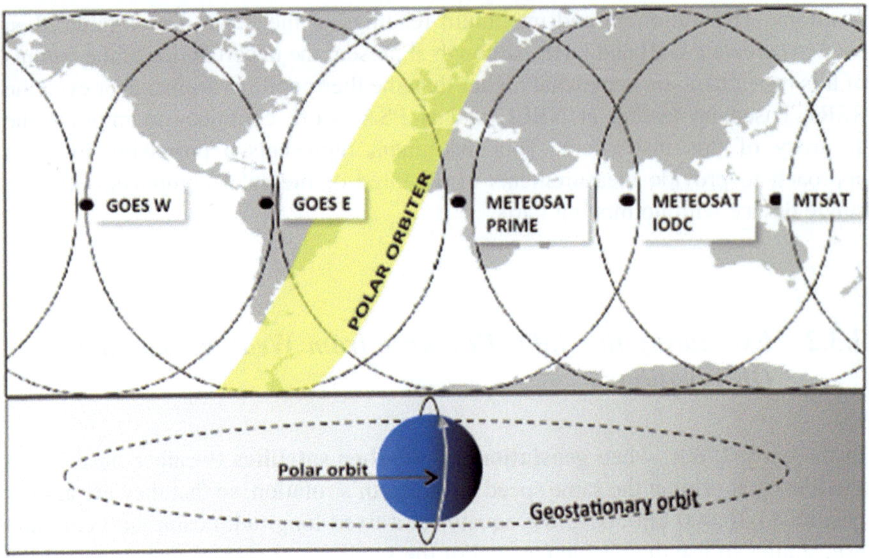

Fig. 3.6 Location of modern-day geostationary satellites (source: Richard Perez, SUNY/Albany)

solar resource calculation can be made; and **physical** methods, where basic radiative transfer theory is applied to the satellite signals to develop a calculation of the solar irradiance at the earth's surface.

The semiempirical approach is the most common method currently used for solar resource assessments, due to its ease of use and also to its ability to retain critical physical principles. Nevertheless, both empirical and physical methods are also in use.

A pioneering study in the semiempirical method was developed by Cano et al. [14]. Many of the modern-day semiempirical approaches are based on Cano's original work. This method has gone through a number of key improvements through the years; however, the basic method is used in well-known products such as the Heliosat Method, the HelioClim, and the German Aerospace Institute's (DLR's) SOLEMI, or Solar Energy Mining product, as well as commercial products produced by companies such as Clean Power Research (SolarAnywhere), SolarGIS, and 3Tier (Vaisala). A complete summary of satellite-derived data sources can be found in Sect. 3.6 of this chapter.

The basic principles of the semiempirical approach were reviewed by Renné et al. [15]. For example, Perez et al. [16] have developed a method following the Cano approach based on the fundamental radiative energy balance equation for the earth-atmosphere system:

$$F^{+}_{TOA} - F^{-}_{TOA} = F^{+}_{G}(1 - \alpha_G) + F_A \tag{3.3}$$

Equation (3.3) states that the net outgoing flux of shortwave radiation at the top of the atmosphere (TOA), or F^{+}_{TOA}–F^{-}_{TOA} minus the amount absorbed by the atmosphere, F_A, is equal to the net flux at the surface, $F^+_G – F^-_G$ or $\alpha_G F^+_G$, where α_G represents the ground albedo.

The main challenge with the semiempirical (and, to some extent, the physical methods as well) is to solve Eq. (3.3) for F^+_G. The main differences among the various methods are the way in which the cloud cover is characterized by the satellite's measurement of F^-_{TOA} and how the forward- and backscattering due to atmospheric particles and the absorption of solar radiation by the various trace gases in the atmosphere, such as ozone and water vapor, is addressed.

For example, under clear sky conditions, an accurate determination of the radiative transfer through the atmosphere that accounts for the absorption as well as forward- and backscattering of the shortwave flux due to haze and trace gases is required. A number of **clear sky** radiative transfer models have been developed over the past two decades to address this fundamental issue. Gueymard [17] provides a comprehensive review of 18 clear sky models and determines that among the most reliable is the REST2. Another model in common use is the simplified broadband version of the SOLIS clear sky model, first published by Ineichen [18]. An overview of many clear sky models can also be found in Sengupta et al. [3].

When clouds are present, the visible downward-looking channel of the satellite is used. Modern-day weather satellites measure this value through a relative scale based on digital counts of the reflected radiation; the higher the counts, the higher the reflectivity of the earth-atmosphere system and, therefore, the higher the cloud cover. For example, the Perez et al. [16] method uses the concept of the cloud index, or C_i, which is determined at the satellite level by monitoring the irradiance measured by the satellite for 15–30 days until a "clear sky" albedo at the top of the atmosphere is determined; this then represents a cloud index of zero. Thus, the higher the value of the irradiance reflected from the earth-atmosphere system, the higher the cloud index, i.e., $C_i = f(F^-_{TOA})$. In the case of the empirical-physical models such as Perez et al. [16], the cloud index is expressed as follows:

$$F^+_G / F^+_{G\text{-Clear}} = 1\text{-}\alpha_{TOA} C_i \tag{3.4}$$

Determination of C_i involves the establishment of the "dynamic range" of pixel counts received by the satellite for any given situation when compared to the values determined at the lowest pixel count, which is assumed to be the clear sky value, $F_{G\text{-Clear}}$ (under which $C_i = 1$). The maximum dynamic range is the difference between the lowest (darkest) pixel counts observed by the satellite (lowest F^-_{TOA} or clear sky conditions) and the highest pixel counts that occur for F^-_{TOA}, which generally occurs under completely overcast conditions. When the pixel count is at its lowest level, clear skies are assumed to prevail ($F^+_G = F^+_{G\text{-Clear}}$ in Eq. (3.10)). Under these conditions, only the clear sky model chosen for the satellite method comes into play. As the pixel counts (cloudiness) increase, the cloud index decreases until $C_i = 0$, in which case it is assumed that complete overcast conditions prevail, and additional models besides the clear sky model also come into play.

A number of factors must be addressed to improve the accuracy of this very basic approach. One is the choice of the clear sky model and its features used in the methodology. Different clear sky models invoke somewhat different approaches in addressing Rayleigh and Mie scattering, especially in the way they handle atmospheric **aerosol optical depth**, or **AOD**, which is a normalized measure of the amount of dust and haze particles in the atmosphere compared to a totally clean, dust-free atmosphere. Obtaining accurate data on the AOD is quite challenging, since there are very few routine global measurements of this parameter. The way the various models address this critical factor represents one of the biggest differences among the various semiempirical and physical approaches, especially with respect to the accuracy of the DNI calculations.

The various ways in which the individual DNI, GHI, and DHI components are calculated using these satellite approaches are also important. Even under totally overcast skies, although DNI is virtually zero, there is some GHI (due to the DHI produced by forward-scattering off of the clouds), so it is important that the models handle the DHI component properly, either using empirical relationships or physical radiative transfer theory.

A third important factor in the semiemperical approach is accurately determining the ground albedo. As noted above the low end of the dynamic range is established by a determination of the lowest amount of reflection from the earth-atmosphere system. However, the presence of snow cover or the reflection from bright sand or water surfaces can result in an erroneous selection of the lower threshold of the dynamic range. Thus it becomes necessary to identify independent data sources to ensure when high reflective values are due to snow or bright surfaces, rather than clouds. Recently some of the semiempirical models are making use of the infrared channels of the satellite to obtain clearer understanding of ground conditions as well as certain cloud characteristics.

Another important factor is the impact that deep, vertical clouds might have on the calculations, since these clouds can cast shadows on adjacent clouds or surface areas that are not properly captured by the cloud index calculations. One approach is to use the infrared channels of the satellites, since cloud temperatures can be determined using these channels. Once cloud temperatures are known, their vertical extent in the atmosphere can be estimated based on upper-air temperature soundings. However, factoring in the three-dimensional characteristics of the clouds as part of a satellite retrieval method for solar resource estimates can involve time-consuming physical radiative transfer theory, so methods for a fast radiative transfer scheme also need to be developed to take advantage of this additional information on cloud characteristics. Work is currently underway at NREL, with assistance from the University of Wisconsin and Colorado State University, to address a more physical-based approach to determining surface solar radiation [3].

In summary, satellite-derived estimates of the solar resource are now in common use. They have the capability of providing monthly accuracies as good as 5 % for GHI and 10 % for DNI, and these accuracies continue to improve with new developments. Satellite estimates also give us the capability of developing solar resource information at a very high spatial resolution and an acceptable temporal resolution (15 min with modern-day satellites).

Section 3.6 summarizes a variety of satellite-based data sets available through public institutions and commercial vendors.

3.3.3 Resource Variability and Its Impact on PV Performance and Grid Flexibility

The temporal variability of the solar resource can have important effects on PV system performance. Since PV systems respond instantaneously to a change in the incoming resource, short-term fluctuations, such as those caused by a cloud passing across the sun, can cause a **ramp** to occur in the solar resource that results in a sudden drop in PV system output. For PV systems that have no internal built-in storage mechanism, understanding these **ramp rates**, which can occur in time frames of seconds to hours, is very important.

The effects of these ramps on the overall output of PV connected to the grid tend to be mitigated when the PV systems are distributed over large areas or even in situations where large-scale ground-mounted systems are only partially shaded by clouds. Under these circumstances, not all of the systems are impacted by cloud shadowing at the same time, resulting in a smaller relative ramp size than is found at any given location. This is exemplified in Fig. 3.7. This figure shows that, at a single location (top panels), fluctuations in the irradiance (left side) and changes in the irradiance (right side) can be large due to cloud passes. When large number of systems are aggregated together, however, (lower panels) the relative magnitude of the irradiance fluctuations and the changes in irradiance for the aggregated systems is significantly reduced [19].

Ramps can be important to utilities over longer time frames, such as over a period of a few hours, where the net load (the difference between the system load and the energy output of variable renewable energy systems such as PV) changes significantly through a portion of the day. In a recent report by the California Independent Systems Operator [20], increased utility "flexibility" will be required to allow for balancing of supply and demand as PV penetrations become greater and greater. This is explained by the **duck curve** shown in Fig. 3.8. The figure represents a hypothetical day (March 31) of diurnal net load patterns for several years in the future where PV penetrations increase each year. Such a scenario results in overgeneration risk due to the high quantity of PV in conjunction with the high midafternoon solar resource and relatively diminished midday electrical load. However, as the afternoon progresses, the solar resource diminishes, yet the load increases associated with the evening increase in electricity usage. Thus a sharp ramp in the net load occurs during the late afternoon hours. In future years this effect will become more and more pronounced as more and more solar is introduced into the system. To maintain reliability under these high-penetration VRE

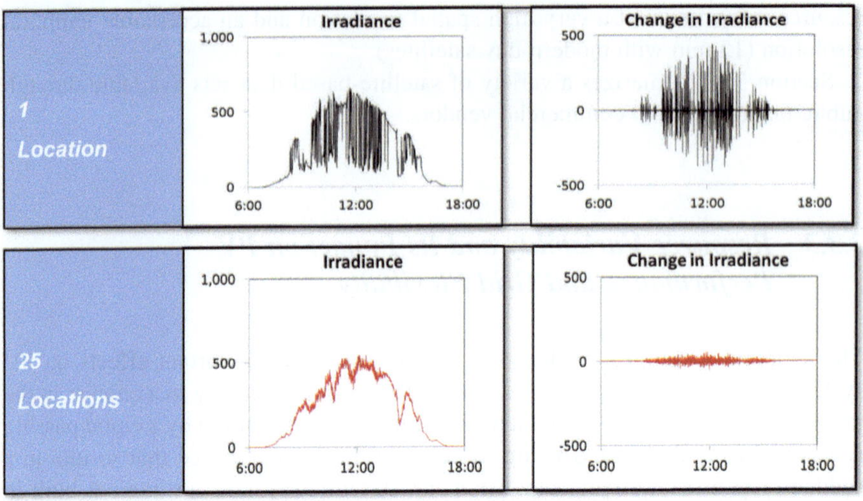

Fig. 3.7 How a 25-station network reduces 10-s variability by over 70 % in a 400 × 400-m area (data from Cordelia Junction, California, USA, 7 November 2010) (source: Clean Power Research and Hoff and Perez [19])

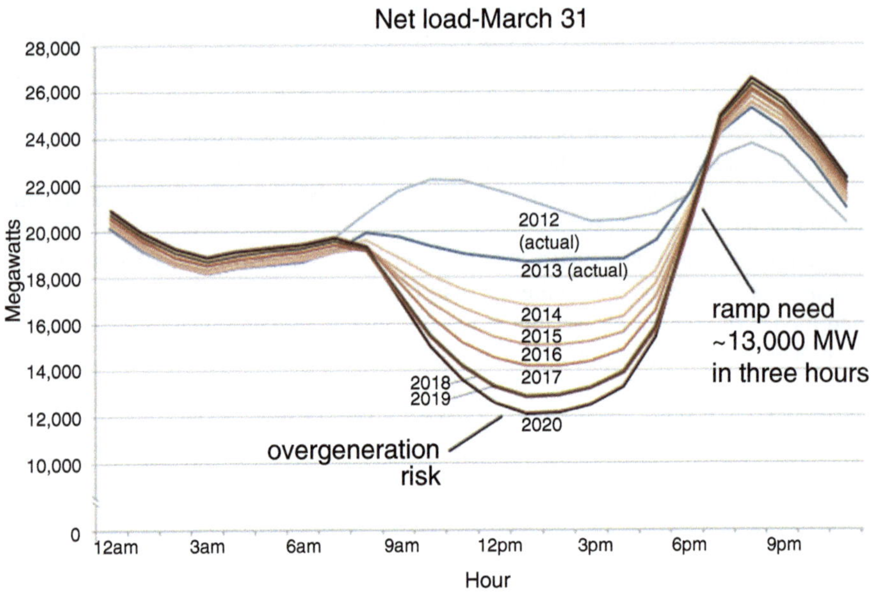

Fig. 3.8 Illustration of steepening ramp rates in the early morning and late afternoons associated with increased penetration of PV into the grid (source: CAISO [20])

scenarios, system operators must have access to resources that have ramping flexibility (including storage technologies), along with the flexibility to start and stop multiple times during the day. Utility systems that already have access to these flexible resources are in a much better position to handle increasing penetrations of variable resources like PV than utilities dominated by inflexible resources such as large-scale fossil or nuclear power plants.

The characteristics of longer-term variability, such as seasonal or interannual, are important for determining how systems might perform over the long run, whether or not they have storage capabilities. This is because most storage technologies serve only to mitigate short-term (sub-hourly) fluctuations, and not those fluctuations that occur from day to day or season to season. Understanding interannual variability can also be important for determining a cash flow analysis of solar systems. Since long-term high-quality ground data sets are seldom available to determine interannual variability at a site, **site adaptation** methods are often necessary, where short-term high-quality ground data are merged with longer-term estimates derived from weather satellites or NWP models to develop reliable information on interannual variability. This approach is described briefly in Sect. 3.4.3 of this chapter.

3.3.4 Predicting the Solar Resource: Forecasting Methods

As more and more solar is installed in national electricity systems, solar resource forecasting and prediction of resource variability are becoming increasingly important for effective utility operations. A number of national and international research programs have been initiated to improve the ability to predict future solar irradiance over different time horizons; for example, a US national solar resource forecasting program led by Dr. Sue Ellen Haupt the National Center for Atmospheric Research [21] and the outputs from the IEA SHC Task 46 activity mentioned in the Introduction. Here we provide a brief overview of the current major focus that is being placed on solar resource forecasting R&D. A comprehensive description of these and more forecasting techniques is available in Kleissl [22].

Resource forecasting R&D typically falls under three categories based on the time horizon of the forecast: short term or minutes to 1 h ahead (sometimes referred to as **nowcasting**) to support load following and balancing, midterm (out to about 6 h) for load balancing, and long term (one to several days ahead) for system planning.

Short-term forecasting research is currently focusing on the use of a ground-based solar monitoring network or one or more **total sky imagers** (TSIs) for observing cloud patterns and movements within the visible horizon of the site. In particular, time-lapse imagery from TSIs to develop minute-by-minute forecasts based on the motion vectors of the clouds observed by the instruments is proving to be a promising tool for solar resource forecasts out to about a 1-h time horizon.

Satellite-derived cloud motion vectors (CMVs) have been studied to predict cloud movements based on satellite-derived time-lapse cloud imagery for midterm cloud forecasting. Results show good success for forecast horizons out to about 5–6 h in many locations, particular where there are no local topographic effects that may produce persistent features such as standing wave clouds.

For longer-term forecasts (one to several days ahead), numerical weather prediction (NWP) models are generally used, sometimes in conjunction with other data sets or as part of machine learning techniques. These models can simulate atmospheric processes in great detail and predict future atmospheric conditions for up to 2 weeks or more. Typical approaches start with the outputs of global- or regional-scale NWP models, such as the European Center for Medium Range Forecasting (ECMWF) model, and then downscale the results using a mesoscale model such as the Weather Research and Forecasting (WRF) model. Traditionally NWPs have not been very accurate in predicting specific cloud types and cloud amounts, so there is currently an active amount of research underway to improve NWPs to predict solar resources with acceptable accuracy for operational purposes.

As part of an international collaboration of the work conducted under IEA SHC Task 46, Lorenz et al. [23] have undertaken solar resource forecasting benchmarking studies. Comparisons of forecasts from a variety of approaches against high-quality ground data were conducted for a number of locations throughout Europe and North America. Figure 3.9 is just one example of many of the results that are available from this collaborative work. The figure compares the results of a number of forecasting approaches, including some that involve combinations of approaches or Model Output Statistics, for a single site and for a large region in Germany for the same time period. The primary metric for the benchmarking shown here is the root mean square error (RMSE) of the forecast result when compared with high-quality ground-based solar measurements (the higher the RMSE, the less reliable the solar forecast). Clearly the forecast reliability is higher when aggregated over a large area rather than at a single site. Further, application of MOS improves forecasting performance. Furthermore, reliabilities generally decrease with increasing time horizons for all approaches. The actual differences among the various approaches are relatively low when compared with the average RMSE for all approaches. Other results besides the example given here show that cloud motion vectors perform better than numerical weather prediction models up to about 3 h.

3.4 Solar Resource Data Sets

3.4.1 Typical Meteorological Year (TMY) Data Sets

Traditionally, typical meteorological year (TMY) data sets have been a common way for assembling a database that can be used to compare the projected performance of solar technologies at different sites. A TMY represents 1 year of hourly

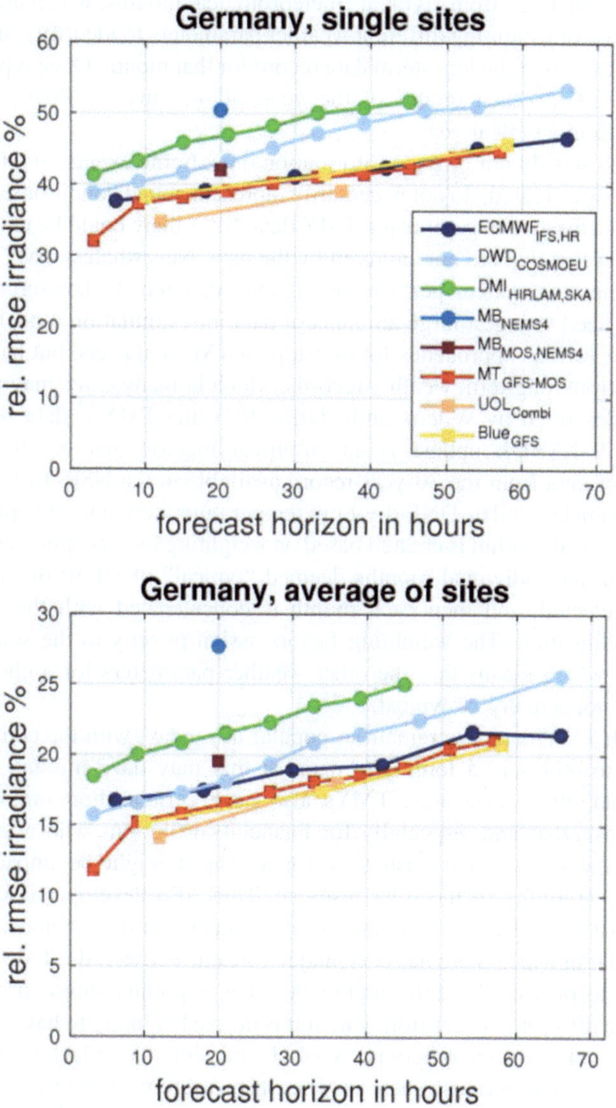

Fig. 3.9 Difference between forecast uncertainty, as measured in RMSE, at a specific location as compared to a large area in Germany for six NWP models, four NWP models using MOS, and two cloud motion vector models (source: Elke Lorenz, [23]). ECMWF, European Centre for Medium Range Weather Forecasting; DWD, German Weather Service; DMI, Danish Meteorological Institute; MT, Meteotest; UOL, University of Oldenburg; Blue, Bluesky Wetteranalyzen

(8760) weather data values extracted from long-term data records (as a minimum, 10 years is generally required). TMY data sets are popular with building designers and others for modeling renewable energy systems, because they provide a relatively concise data set from which to develop system performance estimates. TMY data sets are generated from "typical" meteorological months, which are developed from a process of weighting different weather parameters to identify a single month that is most typical of the long-term data record for that month. Once typical months are established for each month of the year, all months are then concatenated together to form a typical year.

TMY data sets do not provide information on extreme events and, by nature of the way they are created, do not necessarily represent actual conditions at any given time. This is a disadvantage of using TMY data, since there could be unusually high or low solar years that are not captured by the data. Nevertheless TMYs are useful for intercomparing system performance among different technology approaches without the need to ingest large amounts of data into simulation models.

There are various approaches for developing TMY data sets, but basically they are created from long-term weather records, which in many cases may require data filling. As described by Wilcox and Marion [24], the TMY3[2] data set produced from the US NSRDB update is an empirical method that involves selecting 12 months of data from the 30-year record available in the NSRDB based on five weather parameters: GHI, DNI, dry-bulb temperature, dew point temperature, and wind speed. An algorithm is created based on weighting factors applied to these five variables, so that individual months deemed "typical" of all 30 of those months would be selected, and then each month is concatenated with the next month without modification. The weighting factors assign priority to the solar radiation parameters, which means that the other weather parameters for a chosen typical month may not actually be typical.

Although TMY data sets remain in popular use today, with the convenience of being able to represent a long-term data set that may have missing data which required data filling measures, TMYs also have certain shortcomings that are important to understand, especially for financial institutions who need to understand how the year-to-year cash flow for a project might be impacted by the interannual variability of the solar resource. Thus, if a developer is interested in learning what might happen to a system's performance during an unusually cloudy year, or a year in which there has been major volcanic activity, the TMY output will not be able to provide this information. Another important shortcoming of TMY data is that a 30-year data set from which it is derived is likely to have a number of "filled" data values which otherwise would be missing due to lack of observations during certain time periods, such as when a measurement device is down or a station is closed. Thus the TMY value may also not represent the true long-term average value for a given parameter at a given site.

[2] In the USA, Sandia National Laboratories developed the original TMY data set in 1978. When NREL completed the NSRDB in the 1990s, a TMY data set was created out of this 30-year data set and designated as TMY2. An NSRDB update was produced that added in the years 1991–2005, and a new TMY data set was created, designated as TMY3, that makes use of the 1961–1990 data as well as the updated NSRDB data.

3.4.2 *Probability of Exceedance*

To avoid these concerns about the lack of information on extreme events when using TMYs, project designers and developers and financial institutions financing these projects use data sets that provide information on **exceedance probabilities** of the solar resource. For example, a P90[3] value represents the probability that a value will be exceeded 90 % of the time. Exceedance values can be developed from a cumulative distribution function (CDF) of a parameter available from a large (multiyear) data set such as the US NSRDB.

One way to develop a P90 value is to assume that the long-term average data set follows a normal distribution. This can be convenient if the actual distribution of data for the site is not available. However, this assumption adds uncertainty to the analysis and can result in erroneous results. For example, Dobos et al. [25] have looked at the NSRDB GHI and DNI data for Phoenix, AZ, and produced CDFs of these data to illustrate the concept of P50 and P90 (Fig. 3.9). Figure 3.9 shows that, if the annual Phoenix data were fit to a normal distribution (solid line), at CDF = 0.1 (the P90 value), an annual GHI of 2.2 MWh-m^{-2} will be exceeded 90 % of the time (or conversely, the solar resource falls below this value 10 % of the time). Similarly, for DNI, the hourly solar resource exceeds 1.96 MWh-m^{-2} 90 % of the time. However, for Phoenix the long-term solar data does not follow a normalized distribution, so Fig. 3.10 also shows that the P90 value is somewhat lower in Phoenix when determined from an actual vs. a normalized distribution.

A major motivation for developing exceedance probabilities for solar data sets (or for system performance parameters) is to give financial institutions information on the "bankability" of a project. Rather than just knowing what a system output might be under "typical" conditions, exceedance probabilities provide information on the amount of time a system is expected to exceed certain performance levels. Further discussion on data bankability can be found in [26]. The next section discusses how various data sets can be combined to reduce the uncertainty and improve the bankability of the resource data used in project design.

3.4.3 *Merging Data Sets (Site Adaptation) for Reducing Data Uncertainty*

A bankable solar energy project relies on solar resource data with specified uncertainties from which investors can better understand the risk when performance of the system will be lower than desired. Reducing uncertainty in solar resource data is thus a key step toward producing bankable projects.

---

[3] We use P90 here as an example. Other exceedance probabilities, such as P70 and P95, are also of interest to certain project designers and financial organizations.

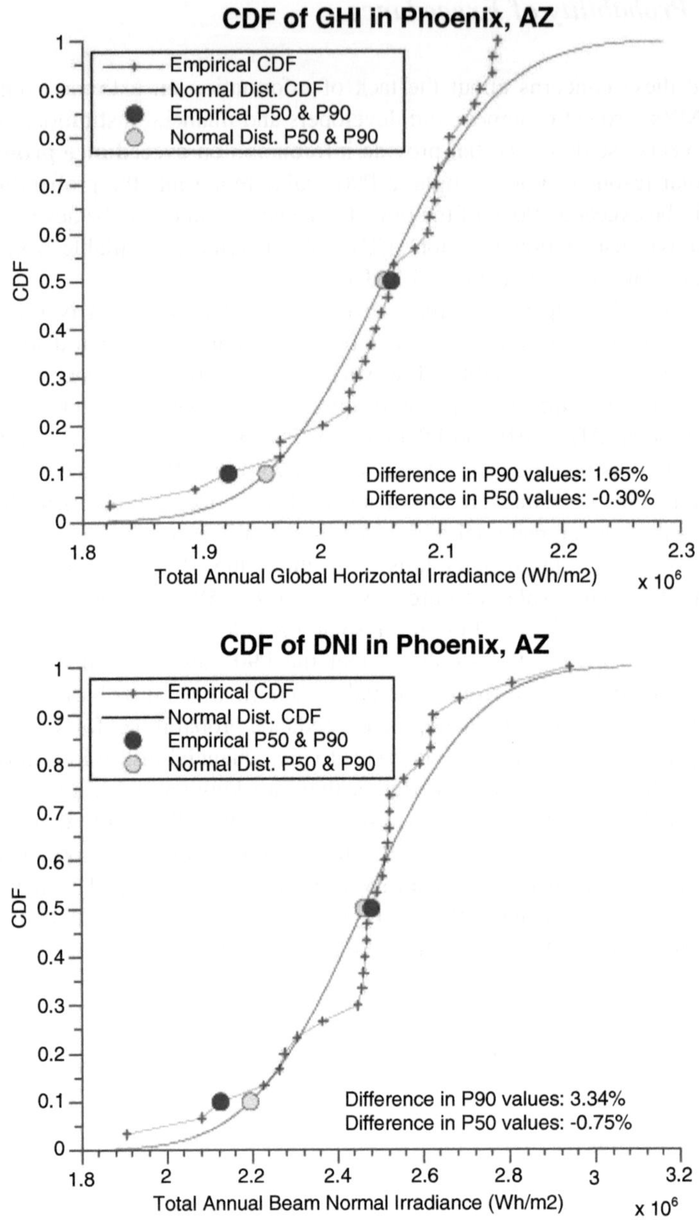

Fig. 3.10 CDF of GHI (*upper*) and DNI (*lower*) for Phoenix, Arizona (USA), as determined from the 30-year NSRDB (from [25])

There are several steps that can be taken to reduce data uncertainty and therefore project risk:

1. Collect high-quality on-site ground data for many years: this is the best way to provide the lowest risk data set to a financial organization; however it is generally impractical due to the time required to collect sufficient data (5–10 years).
2. Collect high-quality on-site ground data for a short time (1 year or less): this would be a more practical approach with respect to investor time scales, but, as indicated in Fig. 3.11, this poses higher risk to investors since they have limited knowledge of interannual variability and solar resource trends and therefore have high uncertainty as to whether the actual measurement period represents a high-resource or a low-resource annual value.
3. Obtain multiyear satellite-derived or other model-derived data sets: in the previous sections, we have noted that we can use ancillary weather data, such as weather satellite imagery or ground observations of cloud cover, to produce long-term data sets where measurements are not available (which is the typical case). This is a relatively inexpensive way to capture enough data to produce P90 values; however, the risk to investors is higher than if actual measured data for the same time period were available.
4. Merge long-term modeled (e.g., satellite-derived) data with short-term high-quality on-site measured data: this approach, known as **data adaptation**, is the most practical and cost-effective approach for developing reliable solar data for a project and offers the lowest risk to developers without causing lengthy project delays.

A detailed description of data adaptation methods is beyond the scope of this chapter; a recent publication by Polo et al. [27] provides a comprehensive review of these methods. The approaches range from a straightforward ratio method to using the CDFs of ground-based data to improve satellite-derived data sets. In the ratio method (e.g., [28]), the ratios of selected periods of concurrent independent data sets (a short-term measurement data set concurrent with the long-term modeled data set) are calculated, and then the ratios are applied to the remaining long-term data set. However, the uncertainties in this approach increase if the biases in either data set vary from time to time. Furthermore, if there are long-term trends in the data, the length of time required of the measured data set increases.

Merging the CDFs of ground-based data to improve satellite-derived data (e.g., [29]) results in a reduction in bias errors and improvement in the Kolmogorov-Smirnov integrals (KSIs) in the satellite data. Mieslinger et al. [30] have recently improved on this approach by determining both the systematic and random deviations of the data sets based on mean bias (for systematic) and standard deviation (for random) statistics and using the KSIs to quantify the differences in the CDFs. Polynomial expressions are then used for the data adaptation. This approach is especially useful when there are large deviations between the satellite-derived and

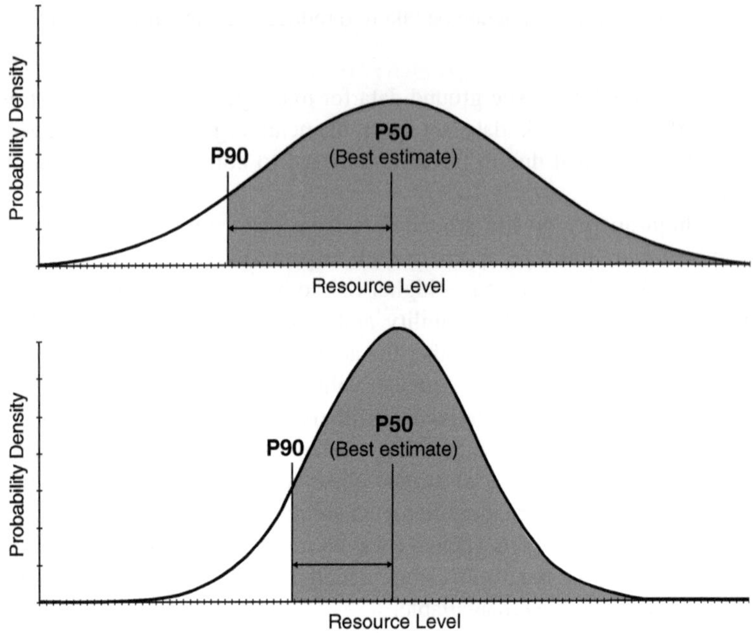

Fig. 3.11 Comparison of data derived solely from satellite methods (*top panel*) with data where the satellite data have been merged with high-quality (lower uncertainty) ground data, even though the period of record for the ground data may be shorter than the satellite data (from [31])

the ground-measured data. The approach requires at least 1 year of overlapping data, but preliminary results show that the corresponding biases approach zero.

Examples of how site adaptation measures can improve the confidence in project performance have been discussed in Schnitzer, Thuman, and Johnson [31]. By examining several case studies, they conclude that uncertainty can be reduced by up to 50 % (from 9.2 to 5.7 %) by conducting quality on-site measurements. The value of this reduction in uncertainty to the bankability of the project is shown in Fig. 3.11, derived from their study. The top panel shows the probability distribution derived from a satellite data set, which has higher uncertainty than quality ground-based data. The effects of merging high-quality ground data with satellite data to reduce the overall uncertainty of the data set are shown in the bottom panel. It can be seen that the best estimate (P50, which could also be obtained by using a TMY data set) does not change between the top and bottom panel; however, the P90 value increases due to the reduced uncertainty. Thus, the merged data set increases the threshold above which 90 % of the data falls, making a project with merged data much more attractive to investors than a project determined from a data set with higher uncertainty.

3.5 Estimating PV Output from Solar Resource Data

Correct usage of appropriate irradiation data is paramount in accurately predicting how much electricity a solar system will generate. As discussed above, poor-quality solar irradiation data for a site location, even when used with excellent simulation programs, will yield inaccurate results when compared with actual system performance. For small projects (<100 kW or so) satellite-derived data may be sufficient for using simulation programs to estimate system output. For larger projects, such as grid-tied utility-scale systems, we have shown that it is best to collect at least a short period of high-quality ground measurements and estimate the interannual variability from the long-term satellite data.

The rating for PV systems applies to system output rating at a resource level of 1000 W/m^2 (1 KW/m^2) at 25 °C (standard temperature conditions, or STC):

$$\mathrm{kW_{peak}} = \text{Area x } 1\ \mathrm{kW/m^2} \times \text{ conversion efficiency} \tag{3.5}$$

The annual energy potential for this condition is then

$$\begin{aligned}(\text{Energy})_{\text{Annual}} = &\ \text{Installed Capacity (kW)} \\ &\times \text{ resource } \left(\mathrm{kWh/m^2\text{-}year}\right)/\mathrm{kW_{peak}}\end{aligned} \tag{3.6}$$

For example, if a system is being deployed that has a total panel area of 4654 m^2, with the panels operating at an efficiency of 17.5 % under STC, the peak installed capacity according to Eq. (3.6) is potentially 814 kW$_{peak}$. However, for a number of reasons, the actual performance of the installed capacity will likely be below this level. This is because a number of other factors must be taken into consideration that can result in a **derating** of a system. When quantified, this means that **derate factors** need to be applied to the estimated system output estimates of Eq. (3.6). Key derating factors are summarized here.

3.5.1 Shading

Shading over portions of a PV array will have the most adverse effect on the system's performance (an optimal shading derate factor is 1.00, and lower factors reduce system output performance). It is important to determine during the site assessment if a potential location for the PV array will be shaded, especially between the hours of 9 a.m. and 3 p.m. This is important as the output of PV modules may be significantly impaired by even a small amount of shading on the array. A careful assessment using an hourly computer simulation program or special instruments is necessary to determine the potential impacts of shading. A minimum of 6 h of unshaded operation is important for best system performance.

Configuring a PV array to avoid shading 100 % of the time is not always possible given space constraints both on building rooftops and on open land. For PV arrays that consist of multiple rows of PV modules and array structures, the losses that occur when one row shades an adjacent row should be accounted for when estimating the performance of the PV system. Steps should be taken in the design of the system to eliminate or minimize these array effects, regardless of the season and solar angle.

3.5.2 Panel Orientation

Next to shading, orientation of the PV array is one of the more important aspects of the site assessment. As discussed in Sect. 3.2, in northern latitudes, tilting the array toward the south at the same angle as the latitude of the site will allow for the direct irradiance from the sun to fall more normally onto the array, thereby capturing more of the resource than if the array is mounted horizontally. Under these conditions, facing the PV array due south is ideal; however, slight deviations can also be feasible. As an example, at a certain site an unshaded PV array with a tilt of 35° and facing ±45° away from due south (SE or SW) could still receive up to 92 % of the annual solar radiation as compared to the PV array facing due south. The orientation can also be optimized by understanding the typical diurnal cloud conditions and the time of maximum load that the array is serving. For example, in sites where clouds tend to be greater in the afternoon than in the morning, tilting the array somewhat east of south will likely receive a higher resource during the morning hours. However, if the maximum load is in the afternoon, it may be better to tilt the array somewhat to the west of south so that, even though clouds might be greater, the resource received by the array can be maximized at a time that coincides with the maximum load.

Based on these considerations, orientation must be understood very early in the project design phase. Often the roof tilt is used as the orientation of a residential rooftop system due to the improved aesthetics of a parallel roof-mounted array rather than an array that is tilted at an angle greater than the roof tilt. Most roof orientations are not the most ideal for the array orientation, so the impact of a less than optimal orientation must be understood prior to solidifying the system orientation.

3.5.3 Reflection

The reflection losses caused by non-perpendicular incoming irradiation are calculated separately for the diffuse and direct irradiation fractions. The direct irradiation is calculated by using the angle of incidence and the level of reflection (being a function of the angle) for a low iron module glazing (standard module types).

Constant reflection losses of 3.5 % are typically assumed for the sum from the direct and diffuse fraction.

3.5.4 *Other Derate Factors*

Several other factors can also impact PV system output. These factors should also be incorporated into the energy production estimates of the proposed system so that system owners can have realistic expectations of how their system will perform and the resulting economic benefits they can expect over time.

1. **Temperature**: PV module output power is reduced as module temperature increases. Thus, temperatures above 25 °C will result in lower output than the DC nameplate rating of the module. The change in output varies based on the electrical characteristics of the specific PV technology, but it can generally be stated that when the module temperature is approximately 25° higher than the ambient temperature, it aggravates the loss in power output.
2. **Soiling** caused by dirt, dust, and bird droppings: These can accumulate on a PV module surface, blocking some of the sunlight and reducing the output. An inclination angle of at least 10° is recommended to allow "self-cleaning" through rainfall, which periodically acts to clean the panels. Practical experience from typical installations show that a 2 % loss on average for a year should still be assumed for soiling conditions.
3. **Mismatch and wiring losses**: There are inconsistencies in performance from one module to another (mismatch) and typically results in at least 2 % loss. There are also losses associated with the resistance in system wiring that can be minimized by increasing the size, and therefore the conductivity, of the wire.
4. **DC to AC inverter conversion** losses: The DC electricity generated by PV modules must be converted to AC electricity to match requirements of common utility loads. This conversion is accomplished using an inverter. Currently, inverters have peak efficiencies of 92–98 % (higher efficiencies are typically associated with larger inverters used in commercial and utility-scale projects).

In summary, the performance of a PV system depends on many variables, and readily available performance modeling tools allow the user to input the amount that the DC output of a PV array will be reduced (or derated) at the point where power is delivered to loads, due to losses in the system. The derate factor, however, should not be used as a hard and fast value for estimating the amount of power or energy that a PV system will generate. When estimating the energy output and savings that a PV system will achieve for a customer, it is best to develop conservative estimates of the system performance or to provide a range of expected output values.

3.6 Sources of Solar Resource Data

This section provides a summary of publicly available sources of solar data, both measured and modeled, as well as key ongoing programs such as the IRENA Global Atlas and the World Bank/ESMAP Resource Mapping Program.

Sengupta et al. [3] provide a comprehensive summary of historical and current sources of data and data products available both through the public domain and through private companies (see Chap. 5.4 of this NREL study). Table 3.1 provides a summary of selected currently available data sets, encompassing both measured and modeled data sources, described in the NREL report.

A key development over the past few years has been the production of the International Renewable Energy Agency (IRENA) Global Atlas for Renewable Energy (globalatlas.irena.org). The Atlas significantly extends an effort first started in 1999 known as the Solar and Wind Energy Resource Assessment (SWERA) project, funded by the Global Environment Facility and managed by the United Nations Environment Programme. SWERA ended in 2007. With the formation of IRENA in 2010 the SWERA data, which resided initially with the UNEP/Global Resource Information Database (GRID) in Sioux Falls, South Dakota (USA), and more recently at NREL (http://maps.nrel.gov/swera) is also available through the Global Atlas. IRENA Global Atlas provides basic foundational information on national renewable energy potentials in order to remove knowledge barriers and stimulate national renewable energy investments and policy formulation. The Global Atlas initially incorporated wind and solar data but is now being expanded to include data of all renewable energy resources.

Another key development launched in October 2012 is the World Bank's Energy Sector Management Assistance Program (ESMAP) solar and wind resource mapping project (http://www.esmap.org/RE_Mapping). This project facilitates renewable energy resource mapping at the country level to support government planning and commercial development. The solar mapping activities include a combination of GIS-based mapping using satellite-derived techniques, with enhancements in some cases provided by numerical weather prediction models. Countries that currently have very limited information on resource potential are targeted for this program, and projects are executed through the World Bank and supported by ESMAP both with funding and technical support. At the time of this writing, preliminary high-resolution solar mapping has been completed for several countries (Pakistan, the Maldives, Malawi, Zambia, and Tanzania), and measurement programs have been initiated to validate the maps for these countries.

3.7 Summary

Although solar resource information has been available in a variety of formats for a number of years, the rapid growth in solar technology deployments, combined with improved ways of both measuring and modeling the solar resource, has resulted in

Table 3.1 Listing of selected currently available solar resource data sources (derived from [3])

Name of data source	URL	Period of record	Coverage	Data elements
World Radiation Data Center (WRDC)	http://wrdc.mgo.rssi.ru	1964–present	Global (>1000 measurement stations)	Primarily daily total GHI; some DNI
European Solar Resource Atlas (ESRA)	http://www.soda-is.com/eng/index.html	1981–1990	Europe, 10-km resolution	Daily and monthly GHI, DHI, DNI
HelioClim (Ecole des Mines de Paris/Armines)	http://www.soda-is.com/eng/index.html	1985–present	Europe, Africa, Mediterranean Basin (5-km resolution)	Hourly and daily GHI
PVGIS (Joint Research Centre, Italy)	http://re.jrc.ec.europa.eu/pvgis/download/download.htm	1981–1990	Europe, Mediterranean Region, Africa, SE Asia; nominally 30-km resolution	Monthly and annual GHI (Europe), daily GHI elsewhere
METEONORM 7.1.4 (Meteotest, Switzerland)	http://meteonorm.com	1991–2010	Global interpolated data from ~8325 stations	GHI, DHI, DNI, other
NASA Surface Meteorology and Solar Energy	https://eosweb.larc.nasa.gov/sse/)	July 1983–June 2005	Global, 100-km resolution	GHI, DHI, DNI
DLR-ISIS (German Aerospace Institute)	http://www.pa.op.dlr.de/ISIS	1984–2004	Global (280-km resolution)	DNI, GHI
Solar Data Warehouse	http://solardatawarehouse.com	Up to 25 years to present	USA (~3000 measurement stations)	GHI
US National Solar Radiation Database updates	www.nsrdb.nrel.gov	1961–2010	USA (1454 stations, mostly modeled)	GHI, DHI, DNI, other
Baseline Surface Radiation Network	www.bsrn.awi.de/en/home/wrmc/	1992–present	Global (~40 measurement stations	GHI, DHI, DNI, other
3Tier by Vaisala	http://www.3tier.com/en/support/resource-maps/	1997–present	Global maps (commercial data sets available)	GHI, DNI
SolarAnywhere (Clean Power Research)	www.cleanpower.com/SolarAnywhere	1998–present	North America (10-km, other commercial data sets available)	GHI, DHI, DNI
GeoModel Solar (now, SolarGIS, Slovakia)	http://geomodel.eu/index.php	April 2004–present	Europe, Africa, and the Middle East (5-km resolution, available commercially)	GHI, DHI, DNI, other

significant activity at regional and national levels to improve and make more accessible the quality and completeness of these data. Furthermore, the many types of products required by government policy makers, project developers, system operators, and financial institutions have been a very strong impetus to expand and improve on the type of information available. This chapter has provided a snapshot of best practices in collecting and using data for PV applications. The reader is encouraged to go to the references and the URL links provided in the text to obtain more comprehensive information about solar resources and how solar resource data can be applied to all solar energy applications.

References

1. REN21 (The Renewable Energy Policy Institute of the 21st Century, Global Status Report (2016), http://www.ren21.net/status-of-renewables/global-status-report/
2. F. Vignola, C. Grover, N. Lemon, A. MacMahan, Building a bankable solar radiation dataset. Sol. Energy **86**, 2218–2229 (2012)
3. M. Sengupta, A. Habte, S. Kurtz, A. Dobos, S. Wilbert, E. Lorenz, T. Stoffel, D. Renné, D. Myers, S. Wilcox, P. Blanc, R. Perez, Best practices handbook for the collection and use of solar resource data for solar energy applications. Technical Report NREL/TP-5D00-63112, National Renewable Energy Laboratory, Golden, Colorado, USA, February 2015
4. R.C. Temps, K.L. Coulson, Solar radiation incident up on slopes of different orientations. Sol. Energy **19**, 179–184 (1977)
5. J.E. Hay, Study of shortwave radiation on non-horizontal surfaces. Report 79-12, Downsview, ON. Atmospheric Environment Service (1979)
6. T.M. Kluchar, Evaluation of models to predict insolation on tilted surfaces. Sol. Energy **23**, 111–114 (1979)
7. B.Y.H. Liu, R.C. Jordan, Daily insolation on surfaces tilted towards equator. ASHRAE Trans. **67**, 526–541 (1961)
8. D.T. Reindl et al., Evaluation of hourly tilted surface radiation models. Sol. Energy **45**, 9–17 (1990)
9. C.A. Gueymard, D. Myers, Validation and ranking methodologies for solar radiation models, in *Modeling solar radiation at the earth's surface*, ed. by V. Badescu (Springer, New York, 2008)
10. W. Marion, S. Wilcox, *Solar radiation data manual for flat-plate and concentrating collectors. NREL/TP-463-5607* (National Renewable Energy Laboratory, Golden, CO, 1994)
11. S. Wilcox, D. Myers, *Evaluation of radiometers in full-time use at the National Renewable Energy Laboratory Solar Radiation Research Laboratory. NREL/TP-550-44627* (National Renewable Energy Laboratory, Golden, CO, 2008)
12. A. Habte, S. Wilcox, T. Stoffel, Evaluation of radiometers deployed at the National Renewable Energy Laboratory's Solar Radiation Research Laboratory. Technical Report NREL/TP-5D00-60896, National Renewable Energy Laboratory, Golden, Colorado (USA), February 2014
13. N. Geuder, M. Hanussek, J. Halle, R. Affolter, S. Wilbert, Comparison of corrections and calibration procedures for rotating shadowband irradiance sensors, SolarPACES Conference. Granada, Spain, 2011
14. D. Cano, J. M. Monget, M. Albuisson, H. Guillard, N. Regas, L. Wald, A method for the determination of the global solar radiation from meteorological satellite data. Sol. Energy **37** (1), 31–39 (1986)

15. S. Renné, David, Richard Perez, Antoine Zelenka, Charles Whitlock, Roberta DiPasquale, in *Use of weather and climate research satellites for estimating solar resources. Chapter 5*, ed. by D. Yogi Goswami, Karl W. Böer. Advances in solar energy, Vol 13 (American Solar Energy Society, Boulder, CO). p. 457 (1999)
16. R. Perez, P. Ineichen, K. Moore, M. Kmiecik, C. Chain, R. George, F. Vignola, A new operational model for satellite-derived irradiances: Description and validation. Sol. Energy **73**(5), 307–317 (2002)
17. C.A. Gueymard, Clear-sky irradiance predictions for solar resource mapping and large-scale applications: Improved validation methodology and detailed performance analysis of 18 broadband radiative models". Sol. Energy **86**, 2145–2169 (2012)
18. P. Ineichen, A broadband simplified version of the SOLIS Clear-Sky Model. Sol. Energy **82**, 758–762 (2008)
19. T.E. Hoff, R. Perez, Modeling PV fleet output variability. Sol. Energy **86**(2012), 2177–2189 (2012)
20. CAISo (California Independent System Operator), 2013: Fast facts: What the duck curve tells us about managing a green grid. 250 Outcropping Way, Folsom, CA 95630. www.caiso.com
21. S.E. Haupt, *A public-private academic partnership to advance solar forecasting. National Solar Conference, 2013* (American Solar Energy Society, Baltimore, MD, 2013)
22. J. Kleissl (ed.), *Solar energy forecasting and resource assessment* (Academic, London, 2013), 504 pp.
23. E. Lorenz, J. Kuehnert, D. Heinemann, K.P. Nielsen, J. Remund, S. C. Mueller, Comparison of Irradiance Forecasts Based on Numerical Weather Prediction Modes with Different Spatio-Temporal Resolutions, Conference Proceedings, 31st EUPVSEC 2015, Hamburg, Germany (2015)
24. S. Wilcox, W. Marion, *Users Manual for TMY3 Data Sets. NREL/TP-581-43156* (Golden, Colorado, USA, National Renewable Energy Laboratory, 2008)
25. A. Dobos, P. Gilman, M. Kasberg et al., P50/P90 Analysis for Solar Energy Systems Using the Systems Analysis Model. Conference Paper NREL/CP-6A20-54488, June 2012. National Renewable Energy Laboratory, Golden, Colorado. Presented at World Renewable Energy Congress, Denver, 2012
26. C.A. Gueymard, A review of validation methodologies and statistical performance indicators for modeled solar radiation data: Towards a better bankability of solar projects. Renew. Sustain. Energy Rev. **39**, 1024–1034 (2014)
27. J. Polo, S. Wilbert, J.A. Ruiz-Arias, R. Meyer, C. Gueymard, M. Suri, L. Martín, T. Mieslinger, P. Blanc, I. Grant, J. Boland, P. Ineichen, J. Remund, R. Escobar, A. Troccoli, M. Sengupta, Integration of ground measurements to model-derived data. IEA-SHC Task 46 Solar Resource Assessment and Forecasting, Final Report for Subtask B3 (2015), To be available on http://task46.iea-shc.org
28. C.A Gueymard, Stephen M. Wilcox, Spatial and temporal variability in the solar resource: Assessing the value of short-term measurements at potential solar power plant sites. Proceedings, Solar 2009, Buffalo, New York, American Solar Energy Society Proceedings, 2009
29. Schumann, Kathrin, Hans Georg Beyer, Kaushal Chhatbar, Richard Meyer, 2011: *Improving satellite-derived solar resource analysis with parallel ground-based measurements*. Proceedings, Solar World Congress, Kassel, Germany, 2011. International Solar Energy Society
30. T. Mieslinger, F. Ament, K. Chhatbar, R. Meyer, A new method for fusion of measured and model-derived solar radiation time-series. Proceedings, SHC 2013: International Conference on Solar Heating and Cooling for Buildings and Industry. Energy Procedia **48**, 1617–1626 (2014)
31. M. Schnitzer, C. Thuman, P. Johnson, *Reducing uncertainty in solar energy estimates* (AWS Truepower, Albany, NY, 2012). 21pp.

Chapter 4
Photovoltaics as a Major Contributor to the Future Global Energy Needs and a 100 % Renewably Powered World

Winfried Hoffmann

List of Abbreviations

APAC	Asia-Pacific (region)
BSW-Solar	German Solar Economy Association (Bundesverband Solarwirtschaft
CAGR	Compound average growth rate
CCS	Carbon capture and storage (also called CSS = carbon sequestration and storage)
CEO	Chief executive officer (chairman of management board of a company)
EEG	Renewable Energy Sources Act (FiT in Germany)
EPIA	European Photovoltaic Industry Association since spring 2015 renamed to SolarPower Europe
FiT	Feed-in tariff
GHG	Greenhouse gases (CO_2, CH_4, water vapor, etc.)
IEA	International Energy Agency
IPCC	Intergovernmental Panel on Climate Change
ITRPV	International Technology Roadmap for Photovoltaics
LCoE(S)	Levelized Cost of Energy or Electricity (Service or Storage)
PV	Photovoltaic
P2G	Power-to-gas ((renewable) electricity → hydrogen by electrolysis → CH_4)
RE(S)	Renewable energy (systems)
RoW	Rest of World
R&D	Research and development

W. Hoffmann (✉)
Applied Solar Expertise - ASE, Bergstrasse 48, 63456 Hanau, Germany
e-mail: winfried@hoffmann-ase.de

A. Sayigh (ed.), *Photovoltaics for Sustainable Electricity and Buildings*,
DOI 10.1007/978-3-319-39280-6_4

SKE	Stein Kohle Einheit (hard coal equivalent)
SME	Small and medium enterprises
UN	United Nations
WBGU	Wissenschaftlicher Beirat der deutschen Bundesregierung für globale Umweltveränderungen (Advisory Board to the German Government on Global Change to the Environment)
WEO	World Energy Outlook (by IEA)
WMO	World Meteorological Organization

4.1 Today's and Future Global Energy Needs

4.1.1 Today's Energy Portfolio

Before answering the future energy needs, it is useful to first take a look of today's energy picture. Figure 4.1 shows the global primary, secondary, and end-user energy needs consumed around 2010. In this paper we will use the term PWh for the various energy numbers which is the acronym for Peta-Watt-hour (Peta $= 10^{15}$ therefore 1 PWh = 1000 billion kWh). The ~140 PWh primary energy in the left column is the energy content contained in the basic raw materials used to produce the secondary energy forms. The data are from IEA [1] and use the "physical energy content method." The major energy carriers for the primary energy (in ~2010) are oil (31 %), gas (22 %), coal (27 %), nuclear (7 %), and all renewables (13 %). It is the pretention of this contribution to demonstrate that renewables could grow toward 100 % by 2050—not only because of ecological but even more so on economical reasons. The second column describes the secondary energy sources with ~90 PWh consisting of treated fossil sources like petrol, diesel, gas, and the convenient energy form electricity. The losses of 50 TWh are mainly associated

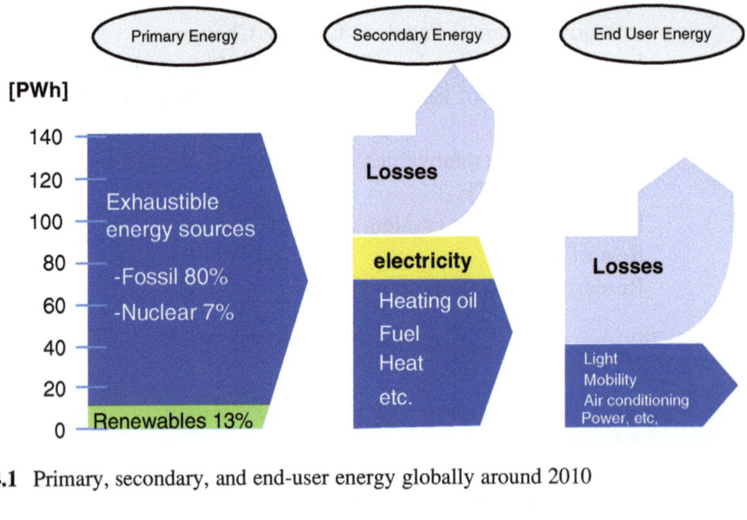

Fig. 4.1 Primary, secondary, and end-user energy globally around 2010

with converting primary energy to electricity in power stations (~30 TWh losses to produce ~20 TWh electricity). The conversion losses of ~50 TWh from secondary energy into what we really need (like light from electricity, transportation with petrol, etc.) are due to the fact that in many cases, we still use highly inefficient products to convert secondary energy into the real customer needs.

The secondary energy is used in various sectors with the following major areas (relative split given for 2010):

- Mobility (27 %)
  which includes passenger vehicles, trucks, aviation and ships, and uses mainly oil
- Low-temperature heat (28 %)
  which is mostly used to heat and cool buildings by using predominantly oil and gas
- Process heat and power for industry (23 %)
- Electricity for industry (9 %)
- Electricity for all other needs (13 %)

The split can be easily remembered as approximately1/4 for mobility, 1/4 low-temperature heat, 1/4 process heat and power for industry, and 1/4 total electricity.

4.1.2 Energy Efficiency Measures

Although energy efficiency measures have nothing to do with renewable technologies, they will nevertheless have a pronounced effect: given an absolute contribution of renewables, the relative percentage will increase by decreasing the necessary secondary energy needs by efficiency improvements. For a better understanding, we will shortly discuss (see Fig. 4.2) three possibilities for decreasing secondary energy while keeping the necessary customer need constant: (1a) use higher efficient products, e.g., replace incandescent light bulbs by solid-state devices, and (1b) displace oil in the mobility sector, e.g., change petrol-driven cars by electrically powered ones, and (2) provide the normally used energy by other technical means, e.g., proper insulation of houses to replace gas and oil for a well air-conditioned home.

1. Proper insulation.
   In many northern countries, a major part of secondary energy is used to provide the gas and oil for a comfortable temperature in private homes, offices, and other buildings. With a well-designed insulation including forced ventilation, it is possible to have even positive energy houses, meaning that, on an annual basis, more energy is produced compared with the need. While this can be accomplished already today cost efficiently with new houses, special emphasis should be on refurbishing existing ones in the years to come. In a country like Germany, the secondary energy needs per year could be decreased by ~40 %.

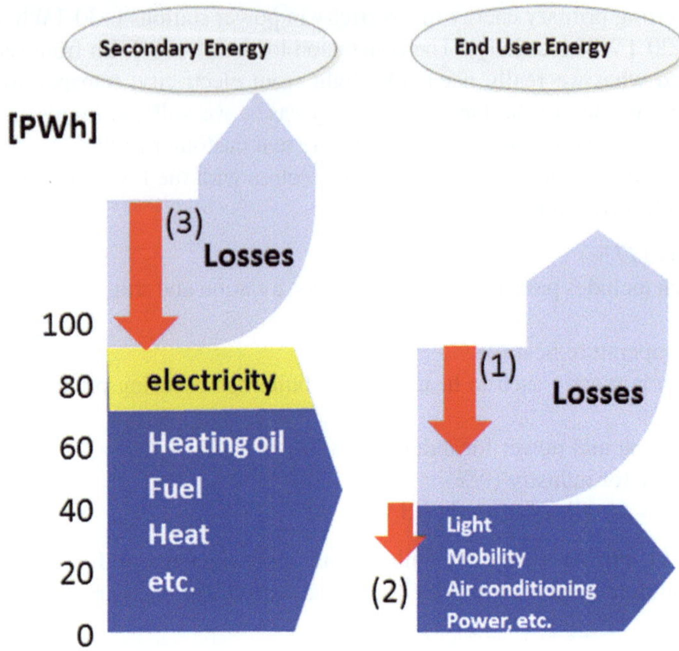

Fig. 4.2 Efficiency measures (*1*) and (*2*) to decrease the needed secondary energy (the use of 100 % renewables will eliminate the need for considering the losses from primary to secondary energy (*3*))

Whatever is still needed for low-temperature heat (including hot water) can be provided by heat pumps and/or solar thermal collectors.

2. Energy-efficient lighting.
An old incandescent light bulb produces the needed light intensity given in lumen at only ~10 lm/W. Modern fluorescent light tubes and solid-state light bulbs like light-emitting diodes (LEDs) and OLEDs (Organic LEDs) are much better with ~(50–90) lm/W. In terms of "same quality of life" (here illumination), the efficiency increase is an impressive factor of 5–9. One important point to mention is the fact that very often the energy-efficient new products are more expensive for the same needs compared to the old ones which is also true in our lighting case to supply a lighting current of ~900 lm: the 75 W incandescent light bulb has a price of ~1.30 €, while the 10 W LED costs ~35 €. However, the important number is "(levelized) cost of service (LCoS)," sometimes also named "total cost of ownership," where we also have to consider the lifetime and the annual running cost of the products. In our example, we have ~1000 h for a light bulb and ~20,000 h for LEDs which leads to ~3.3 €ct/(1000 lm h) for the light bulb and only 0.6 €ct/(1000 lm h) for the LED (assuming 0.30 €/kWh electricity price). If inflation and a price increase for electricity are included, this difference

is even enhanced. Customers have to be well informed about this important fact to make the proper investment decision.

3. Change petrol-driven cars by electrically driven ones.
   It is well known that electrical motors have by far the highest efficiency. In the case of automobiles, we have for the best diesel-powered cars an efficiency of only ~30 %, which can be compared with about 60 % for an electrical car (90 % electrical motor, 66 % for the rest). The increase in efficiency by a factor of 2 is, however, only valid if we compare secondary energy. If the electricity comes from traditional sources, where we have to consider a loss of 2/3 to 1/2 for converting primary energy into electricity, the electrical car has only an efficiency of 20 to 30 % based on primary energy used. Only if electricity is produced from renewables, where by definition the secondary energy equals the primary energy, there is the advantage of electrical cars as stated above.

Based on these examples and only a few more, it is straightforward to conclude an overall efficiency increase by a factor of 2—let us call this a "mild" efficiency increase over the coming decades. By doing so the secondary energy would decrease to 45 PWh, and the split would already be quite different to the one of today: total electricity would significantly increase to ~54 % (31 % electricity for mobility, 11 % for industry, and 12 % for the rest), low-temperature heat ~11 %, and industry (without electricity) ~35 %.

Many more examples for efficiency increase are found in two books by Weizsäcker et al. [2, 3], where the authors conclude that globally an efficiency increase in total by a factor of 4 and even up to 5 could be envisioned—the latter may be called an "aggressive" and "very aggressive" efficiency increase. In the following we will consider a value in between of the two cases (mild and aggressive) above, namely a factor of 3, which may be called "realistic" efficiency increase.

4.1.3 Future Secondary Energy Needs for Ten Billion People with "Same Energy for Everyone"

Two important factors have to be considered when discussing the future energy needs: (a) how many people will be on our planet in the future and (b) how could we solve the inequity of today's energy distribution for future generations.

1. Development of future global population
   In its recent document, the UN [4] concluded in their medium scenario that most probably the steep increase in the last century from ~1 billion in the 1920s to 6 billion in the 1990s will be greatly retarded in the coming decades. While the average increase for each decade was a steady increase from 20 to 85 million in the last century, it was already decreased to 75 in the first decade of the new century and is expected to go down to 30 million by 2050. Overall there seems to

be good evidence that the future global population will level off at around 10 billion people.

2. Solving today's inequity of energy distribution
   A simplified calculation for the unfair energy distribution at the turn of the millennium is as follows: from the ~6 billion people globally, using 90 PWh secondary energy, only 1/4 of them used 3/4 of the energy. This gives us the "high-quality" standard, as we in the Organisation for Economic Cooperation and Development (OECD) countries are used to a number of 45 PWh per 1 billion people. Hence with today's energy efficiency, the future 10 billion people would need 450 PWh secondary energy, if all have the high-quality standard.

Taking into account the efficiency measures described before and using the "realistic" efficiency factor of 3, the challenge for the future will be the delivery of 150 PWh secondary energy by renewable sources.

4.2 Necessity to Change from Traditional Fossil/Nuclear to Renewable Technologies

4.2.1 Global Warming

Despite the attempt from a few but well-financed lobby groups and think tanks who are trying to find reasons for the denial of the temperature increase mainly caused by anthropogenic CO_2 increase, there is globally a widespread belief and understanding that we really have a severe problem which should be tackled as soon as possible. This is not only shared by the IPCC community (Intergovernmental Panel on Climate Change) but also from most scientific institutions worldwide. The melting of glaciers everywhere; the occurrence of not more but heavier tornados, hurricanes, and tropical cyclones; the steady increase of the sea level; and the increase of droughts and flooding are indicative of a severe change in the global climate. Fortunately, on a globally political level, 195 countries adopted in Paris on 12 December 2015 an agreement to "holding the global average temperature to well below 2 °C above pre-industrial levels and to pursue efforts to limit the temperature increase to 1.5 °C". The ratification of this world's first cpmprehensive climate agreement is foreseen by 4 November 2016.

The increasing temperature is also resulting in a worrying thawing of the permafrost resulting in a considerable release of methane which has a much higher global warming potential (~twenty-five times greater than CO_2 in the short term). By combining the anthropogenic CO_2 increase and the additional CH_4 as described above, it is highly questionable whether the often discussed upper limit of a +2 °C increase could ever be reached—even if we stopped immediately the further release of CO_2. Given the reluctance of many countries to take serious actions against further burning of fossil fuels as quickly as possible, it is unfortunate but realistic to see a temperature rise of +4 °C and even more with catastrophic impacts on climate, sea level increase, and related consequences (acidification of sea water, change in sea currents like the gulf stream, and many more). In short, it is not "5 min before 12" but rather "2 min after 12" already! It is frightening that many still do not see the urgency to change our way of providing energy from traditional nuclear and fossil to renewables now.

4.2.2 Cost Issues for "Clean" Fossil (CSS)

Given the urgency to cut CO_2 emissions as described above, it is evident that many attempts have been undertaken to explore the separation of CO_2 from the exhaust gas from fossil electricity-generating power stations. Looking to the most recent energy forecasts from oil companies (e.g., BP [5]), it is no surprise that the future use of fossil fuel (coal, gas, and oil) is assumed to take place with so-called CSS (carbon sequestration and storage). While the separation of CO_2 is "only" a technical and financial issue, the ultimate storage of the gas remains as a principal problem. Even if the latter could be solved, there is widespread evidence that a "clean" kWh electricity by fossil power plants would have a cost of > 10 €ct—which is more than electricity produced by PV and wind today. The hype with shale gas does not also help: although a gas-fired plant emits only ~ half of CO_2 for each produced kWh compared to a coal-fired one, the necessary CSS would also add significantly to the "clean" electricity production. It is a shame to observe that these facts are suppressed in the public, political, and industrial discussion just and because incumbent industries are only looking for a short-term profit: without obligatory CSS, it is still rather cheap today to burn fossil fuels and burden future generations with a resulting utterly devastating climate change.

4.2.3 Cost and Safety Issues for New Nuclear

Much has been discussed in recent years after the two accidents in Chernobyl in Russia (1986) and Fukushima in Japan (2011) regarding safety issues and how to avoid such disasters. Whether this risk plus the unknown cost for safe ultimate storage of nuclear waste should be balanced against carbon-free production of electricity is today no longer necessary in order to decide against nuclear. This is because latest contracts have demonstrated that new nuclear is significantly more expensive compared to today's PV and wind electricity production. Great Britain has negotiated a contract with French authorities that they would pay a fixed price of ~11 €ct for every produced kWh from two new nuclear power stations at Hinkley Point. This payment is fixed for 35 years with an additional increase according to annual inflation. As English business people are known for their scrutinized negotiations, this payment seems to mirror the true cost (plus margin) for new nuclear.

Given the fact that new PV installations in Chile produce (with decent profit) electricity well below 3 €ct/kWh and even in Germany with ~7 €ct/kWh and wind onshore below 7 €ct/kWh, it is evident that simply the cost argument rules out new nuclear everywhere. At the end of this book chapter, it will be shown that the LCoE (levelized cost of electricity) for the then new PV and wind systems will become comparable to fossil and nuclear kWhs within the coming decade even including storage.

4.3 Portfolio of "Sustainable" Renewable Energy Offering

In Fig. 4.3 an illustrative comparison is made between the global energy offer for exhaustible primary energies (four small boxes on the right), the annual solar radiation to the continents (big yellow box), the annual primary energy needs (small red box in the middle), and the technical potential for the annual energy

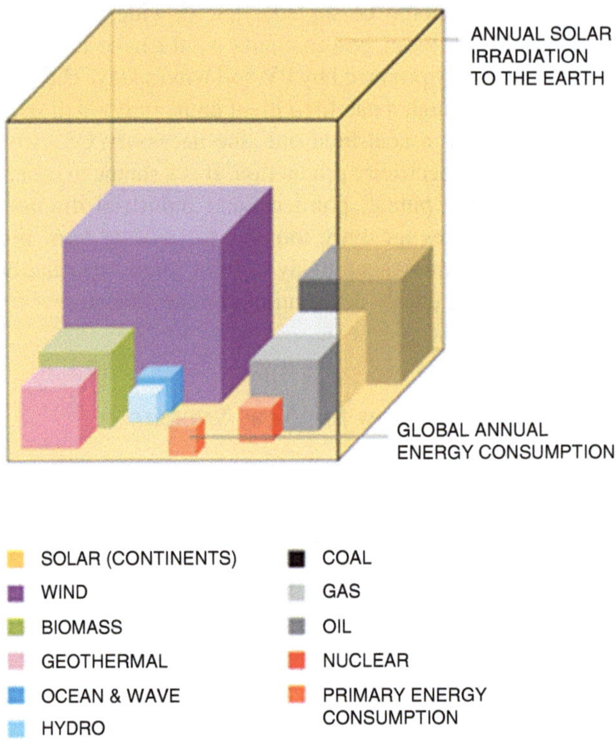

Fig. 4.3 Worldwide energy offer for exhaustible primary energies and annual solar radiation in comparison to the annual primary energy needs and technical potential for the annual energy from various other renewables (source EPIA/SPE)

from various other renewables (small boxes on the left). It is always astonishing to see how dominant the energy based on the annual solar radiation, which even only the continents receive from the sun, really is. The boxes for exhaustible energies are only considering the conventional reserves. The other renewable sources are much smaller compared to solar and are based on their technical potential. From WBGU [6] an effort was made to transfer the technical potential to a so-called sustainable potential; in short the following boundary conditions were applied for the individual renewable technologies:

- **Biomass** should not compete with food production and biodiversity should be preserved; in addition the much higher land use which is one to two orders of magnitude higher compared to PV and wind for the same energy produced should be considered.
- **Geothermal** for electricity production is restricted to those regions which supply cost efficiently high-temperature steam.
- **Hydro** should take into account the methane emissions from flooded areas, the siltation of large reservoirs, and the impact on biodiversity.
- **Wind** is restricted by noise emissions near to buildings for onshore farms.
- **Solar** has limitations when solar farms are competing with food production—dual land use to grow vegetables beneath the solar modules decrease these limitations.

Table 4.1 summarizes the findings from WBGU for the technical and sustainable potential for the various renewable energies. The sustainable potential shrinks by a factor of 25 compared to its technical potential. Fortunately, when comparing to the future needed 150 PWh per year, there is still 21 times more sustainable offering available; these numbers only consider energy offering on continents (and near the coastline for offshore wind). This clearly shows that the often used argument "not enough space available for solar and wind" is not valid.

Figure 4.4 summarizes in the first two columns impressively these findings for the technical (80,000 PWh) and sustainable (3100 PWh) renewable energy offerings on a relative scale. In addition the pragmatic split for a future 150 PWh renewable energy supply is shown in the third column and discussed in more detail in the last chapter.

Table 4.1 Global potentials for renewable energy sources (for comparison: global primary energy consumption in 2010 ~ 140 PWh)

	Technical potential [PWh/year]	Sustainable potential [PWh/year]
Biomass	224	28.0
Geothermal	202	6.2
Hydro	45	3.4
Solar	78,400	2,800.0
Wind	476	280.0
Total	79,347	3,117.6

Source: WBGU Flagship Report [6]

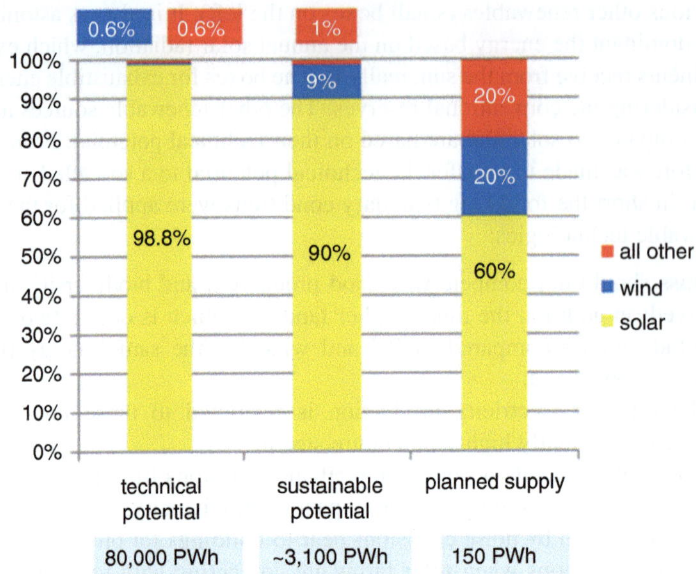

Fig. 4.4 Technical and sustainable renewable offerings per year [6] and pragmatic split for the needed 150 PWh secondary energy needs for the future

4.4 PV Market and Technology

4.4.1 History and Status

Only a few years after the discovery of solar cells in 1953 [7], their economic advantage was utilized to power satellites instead of using more expensive atomic batteries. The first satellite was Vanguard 1, launched in 1958, which only needed a few dm^2 solar cells to power the needs for this rather small satellite. This may be compared to today's largest system, the International Space Lab, which has an area of 1,632 m^2 of solar panels with a DC power of ~130 kW.

The further development very quickly demonstrated that a variety of different customer needs could be powered cost effectively by PV systems. This development is summarized in Fig. 4.5 and highlights, where appropriate, the market size in the beginning as well as the CAGR (compound average growth rate) of the respective decades.

The 1970s saw a multitude of off-grid systems to be cost efficiently powered by PV. This included cathodic protection of pipelines, remote power for transmitter stations, buoy, solar home systems for basic power supply in developing countries and many more.

The advent of thin-film PV modules in the 1980s triggered the powering of many consumer appliances. Like in millions of calculators, it was sufficient to replace the

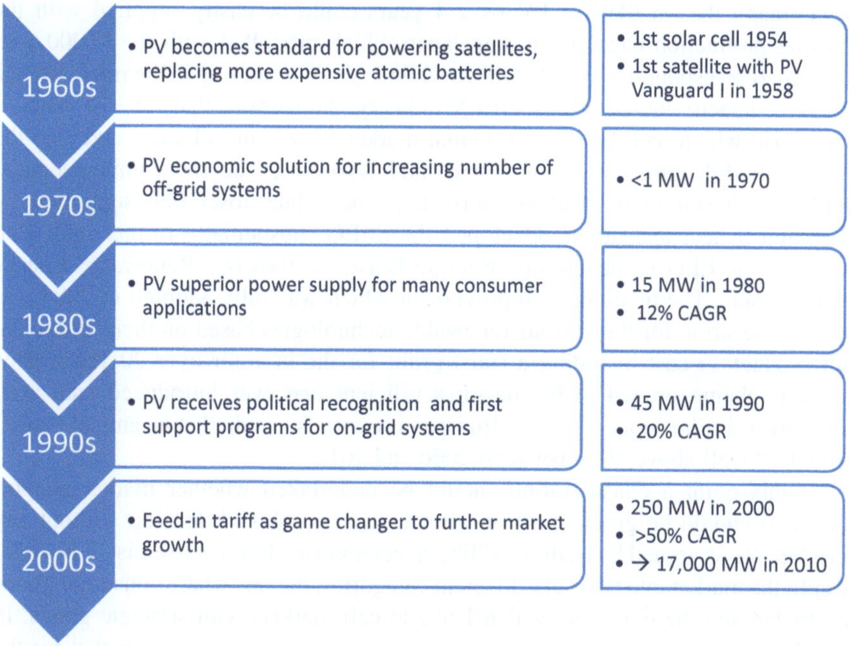

Fig. 4.5 Major PV applications in the respective decade

button cells with a few cm^2 of a thin-film module, mainly consisting of amorphous silicon cells.

It was in the beginning of the 1990s when in Switzerland and Germany, only a few individuals pushed support programs for the introduction of on-grid systems in family houses. In Switzerland it was Hans Ulrich Flückiger who pushed with local political support for a remuneration of 1 Swiss franc for each produced kWh from a PV system. In Germany it was Wolf von Fabeck who advocated for the first feed-in tariff with adequate profit within the lifetime of the invested PV system. This was paralleled by the German "Stromeinspeisegesetz" in 1991 (StrEG = electricity feed-in law for renewables). The resulting "1000 roof program" in Germany, which grew to ~2,500 PV systems (due to Germany's reunification) with an average power of ~2.5 kW each, was deployed during 2001 and 2004 and was the first demonstration that well dispersed in an industrialized country like Germany—even with its limited solar radiation of ~1,300 kWh/m^2—a substantial fraction of the annual electricity needs for a family house could be produced with such a PV system. In addition a major cornerstone was laid from a regulative point of view with respect to a fixed payment over 20 years and the priority of grid access for all produced renewable kWhs. Unfortunately one important point was missing: the combination of a politically induced substantial market volume and a parallel-supported industrial policy for appropriate capacity increase for manufacturing of PV components (modules, inverters, materials, and production machinery). While in Germany the ~6 MW of PV over 4 years could be easily supplied with the existing production capacity, it was Japan which propelled with its 70,000 roof program (corresponding to ~210 MW installed systems) starting in the mid-1990s, a massive capacity increase for their PV industry. No surprise that for ~10 years, it was Japan which had the biggest annual market for PV installations and also the biggest module producers (e.g., Sharp and Kyocera). The most unprecedented market push started in 2001 in Germany, when Hans-Josef Fell supported by Hermann Scheer and others developed the StrEG substantially further into what is known as EEG (Erneuerbares Energie Einspeise Gesetz = Renewable Energy Sources Act). The qualitative improvement which was fully adopted in 2004 was the remuneration for the various renewable technologies based on their individual full production cost including a (fair) profit for the investor over 20 years. This support scheme proved to be the most efficient one ever introduced and, after adopting a similar support in most European countries, resulted in an annual market growth of well above 50 % between 2000 and 2012.

At this point a consideration should be undertaken whether market support programs should be given at all or whether markets should develop without any interference as argued by traditional liberal economists. For this exercise it is useful to split the market into two very different categories: the one where consumer goods are traded and the other one which I like to call markets with strategic goods. In Fig. 4.6 a few examples for the two categories are given. It is obvious that for the

„*Consumer Goods*" like

- *mobiles*
- *laptops*
- *FPD television*
- *automobiles and many more*

only need a global fair trade: customers simply *want* the new product

„*Strategic Goods*" like

- *the way electricity is produced*
- *the way transportation is done*
- *the way people live in the urban environment*

do need initial political and public support

Fig. 4.6 Strategic and consumer goods in a national economy first one, the traditional way of market development should be followed: products are developed and produced globally in free competition and fair trade. The customers simply want to have the next generation of those consumer goods.

In contrast, strategic goods would not develop properly in an undisturbed market environment. Few examples are the following:

1. The catalyst for automobiles was introduced in Japan, the USA, and Europe not because the car industry wanted to offer a better automobile but because of a sociopolitical decision in those countries for a better well-being.
2. The desulfurization and denitrification of waste gas from fossil electricity power stations (SO_x and NO_x removal) were not offered from utilities as a more environmentally friendly electricity, but it was again the sociopolitical decision for a cleaner world.
3. In the 1960s, it was a worldwide accepted belief that nuclear power would be the most cost-effective technology to produce electricity; however, in the beginning of this new technology at low volume, the cost to produce a kWh was much higher compared to the conventional power stations at that time—only the socioeconomical decision to support billions of dollars made it possible that nuclear power could increase its capacity (by today we know it better).

The annual global market growth is shown in Fig. 4.7, and the dominance of European countries until 2011 is clearly seen. From 2000 to 2010, the PV market had an outstanding annual growth of more than 50 %.

Parallel to this market growth, there was also a growth of European production companies like RWE Solar (later SCHOTT Solar), SolarWorld, Q Cells, and some others which had amidst the first decade a significant market share.

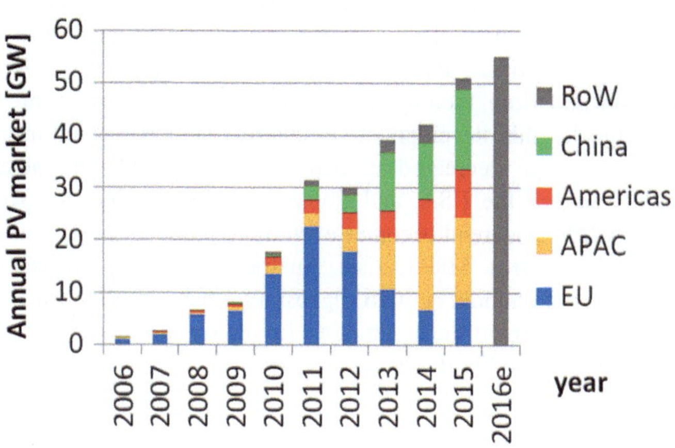

Fig. 4.7 Annual grid connected and other installed PV systems for the major regions (EPIA/ SPE [8])

This picture changed dramatically when - first not well noticed - in the second half of the first decade, a number of Chinese companies increased significantly their production capacity. From the authors perspective, this development was driven by the following boundary conditions:

1. After a thorough analysis of what could be important future global markets and technologies, the Chinese government concluded that PV may become such a candidate, and the Chinese industry should become the world leader for the needed components.
2. The high market growth in Europe asked desperately for more products, so there was no need at that time to take the burden for additional market volume in China.
3. With massive financial support from the Chinese state banks (long-term, low-interest rate), many production companies established along the value chain (cells and modules, crystal and wafer, polysilicon) with many of them also successfully going public at the New York Stock Exchange.

An earlier study by Deutsche Bank Research concluded that the credit volume offered to the Chinese companies was well above 40 billion US dollars.

It should be highlighted that the massive capacity increase was successfully done with production machines mainly from European and US suppliers. These multibillion-dollar turnover numbers for the machine industry definitely helped these economies in the difficult times after the Lehman crisis. However, the massive capacity increase which was for years even significantly higher than the market growth resulted in 2011/2012 in an overcapacity of almost 100 %. This had a very positive, but also very negative, effect: as with every industry which has a substantial overcapacity, the prices are heavily declining and often falling below production cost. This exactly happened for virtually all PV production companies in 2011/2012 where the losses for the big players climbed up toward many hundred million US dollars for each of them. This negative effect has of course also a positive element as the prices for the PV systems allowed new applications which could never be envisioned before, thereby opening new markets in many more countries. Looking to Fig. 4.7 this development is clearly seen: while unfortunately the installations in Europe decreased dramatically after 2011, the other regions APAC (mainly Japan), the USA, and China increased significantly their annual installations. In 2015 this trend becomes even more pronounced with new entrants like India, South America, Middle East, Australia, and some African countries. Market forecasts in the middle of 2016 suggest a further growth toward ~55 GW at the end of 2016 and support the assumption for an average annual growth for this running decade (2010–2020) of ~20 %.

4.4.2 *Price Experience Curve for PV Modules with Resulting System Price*

"Cost learning curves" are well known in the production industry and have demonstrated for a variety of different products that if one plots on a double logarithmic scale the cost per piece against the cumulated number of produced pieces, a straight line is obtained. From the slope the so-called experience factor can be calculated which states that for each doubling of the cumulative production, the cost decreases each time with a characteristic %-number. The Boston Consulting Group used the same method to characterize the *price* development for a variety of products. The resulting graph is called a "*price* experience curve (PEC)," and from the slope, the "*price* experience factor (PEF)" can be calculated (sometimes also called Learning Rate, LR). As an example the price per bit for DRAMs followed for many decades a PEC with a PEF of ~40 %. In absolute terms the price per bit decreased from ~1 mio $ per GB in the 1970s toward only ~1 $ per GB today. The price per m^2 of flat panel displays decreases with a PEF of 35 % [9]. Paul Maycock was the first to introduce this method in the 1990s also to the price development for PV modules with a PEF of ~20 %. The latest PEC update for PV is seen in Fig. 4.8 [10]. For reasons of comparison, the prices are taken for "power" modules only based on crystalline silicon. Constituents for the reduction of the production cost were a reduction of specific material costs, increase of efficiency, and "economy of scale" in

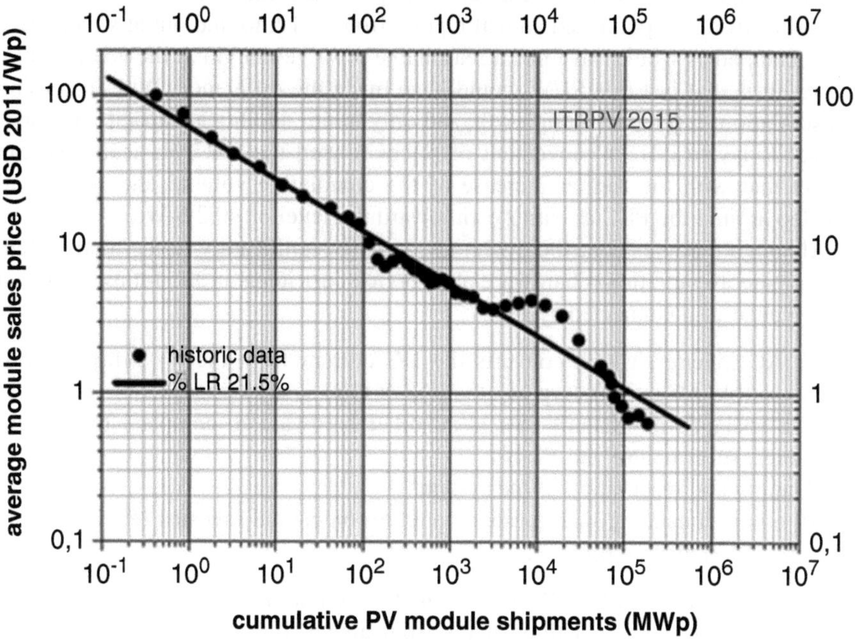

Fig. 4.8 Price experience curve for PV modules (ITRPV 2015 [10])

production. This is exemplified by the specific silicon material cost: the wafer thickness was reduced from 0.7 to 0.15 mm and the kerf loss from 0.5 mm (diamond inside hole saw) to 0.1 mm (wire saw); the material usage per dm^2 could thereby be reduced from ~28 g down to 6 g. Together with the reduction of the price in the chemical industry for polysilicon from ~60 to ~20 $/kg, the material cost per dm^2 was reduced from 168 to 12 $ct. The increase of efficiency from 8 up to ~20 % resulted then in the specific polysilicon material cost from 210 $ct/W in the 1980s to only 6 $ct/W today—this is a reduction by a factor 35! During this time the annual throughput of a single production line for cell and module production was increased from 1 MW (manufactory) to 200 MW (fully automated with high yield).

Sometimes the real price points deviate from the straight line which may have many different reasons. The increasing prices in the years corresponding to a cumulative volume of ~10 GW can be attributed to a shortage in the base material polysilicon and to the market demand created in many countries. As the bottleneck was afterwards removed, we not only quickly reached the PEC curve again but even see an undershoot. This phenomenon is also known from the semiconductor industry with the so-called pig cycles. At a cumulative volume of 100 GW, there have been market prices of around $0.8/W. This is considerably below what one would have expected from the PEC (~$1/W). There is—unfortunately—proven evidence that in those days, we had unhealthy prices, proven by the fact that all major module production companies showed deep red figures in their 2012 annual reports. This was due to an enormous overcapacity in the producing industry and a slowdown of the market growth causing heavy competition and thereby resulting in the low prices observed. Interestingly enough is the fact that if one calculates backward from the published annual reports from 2012, to find out at which price those companies would have had a reasonable positive margin at the volume sold, the result is that just this $1/W would have made the difference. With increasing consolidation, further market growth, and only minor additional capacity increase, we see today in 2015 a movement of prices toward the PEC. Basically we can expect that with the further increase of the cumulative volume, the prices will develop around the PEC at least down toward the level of ~0.2 $/W.

Although modules contribute with about 50 % significantly to the total system price, there are additional hardware and service cost items in the following given for a typical grid connected system:

1. DC/AC inverter, which converts the DC current from the PV array to grid-compatible AC current (typically ~10 % of total system price).
2. BOS (balance of system) cost is all other hardware to complete the system (cables, substructure, etc.).
3. Planning and installation of the system.
4. All others (financing with reasonable return on capital invested (ROIC), insurance, O&M, dismantling, etc.).

A typical price for a MW PV system in 2015 is at around 1€/W and contains the module price and the first three items from above. If we assign, for simplicity reasons, an additional 1€ for the item #4, we can in a first approximation calculate the LCoE (levelized cost of electricity) as follows:

- Electricity produced over 20 years in a place like Germany (1 kWh/(year and $W_{PV\ installed}$) = 20 kWh.
- Total cost 2 €/W.
- LCoE = 200 €ct/20 kWh = 10 €ct/kWh.

In a recent tender offer in Germany for greenfield systems, companies were awarded the contract with ~9 €ct/kWh (which even contained a (small) margin). Obviously this simple calculation gives a reasonably good number. In places with twice as much sunshine (North Africa, Arizona, etc.) the LCoE bisects to ~5 €ct/kWh (if all other parameters stay the same; very often the financing cost is higher in a number of Southern countries and offsets—at least partially—the increase in solar radiation).

Small rooftop systems of several kW have a typical price of ~1.5 €/W. If we make the same simple calculation for a German system, we arrive at ~15 €ct/W, which is well below the price, which home owners have to pay to their municipality (~28 €ct/kWh).

This has often been called the first Holy Grail of "grid parity" and is now obtained already in many countries. While this "grid parity" was not really a fully competitive price, we are now moving toward the second—or real—Holy Grail of "consumer electricity cost parity" (including storage cost for a PV-produced kWh on demand).

4.4.3 *Future Development of the PV Sector*

4.4.3.1 Future PV Market

Today's annual volume for the PV market allows—as with every new industry which gets mature—an appropriate differentiation for the various products to serve specifically the variety of different customer needs. In Table 4.2 the future portfolio

Table 4.2 Differentiated system portfolio powered by PV

Type	Description of specific customer need	PV power	...results in
Off-grid	Consumer appliances	~W	Smart appliances
	Pico-, μ-solar home systems: light, education, communication, water, etc.	W ... MW	Smart power for basic needs
	PV-diesel battery	kW ... MW	Smart industrial power
On-grid	Residential, multifamily houses	kW ... ×10 kW	Smart homes
	Commercial (offices, supermarkets, hotels)	×10 kW ... ×10 MW	Smart business
	Utility (energy regions combining rural and urban areas)	MW ... ×GW	Smart municipality/city/region

for the multitude of PV systems is summarized. Although "smart" has developed in recent years into a buzzword, it seems appropriate to use it for the multitude of PV-powered products and solutions.

Off-grid PV systems have been economically successful since decades as already described, however, only in small niches with a small volume. Originally the powering of small *smart appliances*, like calculators and garden lights, was in the range of a few mW. With the advent of new products like flexible PV, OPV (organic PV), and DSC (dye solar cells) products, aimed to serve flexible surfaces and nicely fitting colors, this type of application can be increased enormously. In the future we will have PV products integrated into tents, rucksacks, and many more surfaces.

Thanks to the parallel development of efficiency improvements and price decrease for PV and batteries, there is now the opportunity for a large volume of powering the *smart basic needs* for the billions of people who otherwise would not have access to energy for a long time. I remember when I was planning and installing solar home systems in Indonesia in the 1980s which consisted of three light bulbs (25 W each), a 300 Wh battery with charge controller, and a 50 W PV panel. The price for the hardware as described was ~500 €. The same service today, namely 4 h of light, can now be provided with much higher efficient LED light bulbs (5 W each), a much smaller 60 Wh battery which results in a significantly smaller PV module of 10 W. The smaller energy needs together with the decreased unit cost for the components give a price for the same customer need which is about a factor of 10 smaller. In addition, when I worked on such projects in the 1980s, the financing had to come from ministries like BMZ (German ministry for technical cooperation) and/or the World Bank. Fortunately and thanks to the Grameen Bank founder, Muhammad Yunus, the possibility for micro-credits to be asked and paid for by the people who need the systems has in the meantime become a standard worldwide. Starting today with even very small pico-systems to recharge the mobile and have one small light source, the powering of all these smart basic needs will with time grow toward powering small villages with all their needs including clean water supply, school, hospital, and small industry powering.

Another important application is the area of *smart industrial power*. This is no longer the earlier very remote powering of repeater stations on mountains far away and only reached by helicopters. But now PV systems are added to large existing MW-diesel generators, e.g., for mining companies, which results in a decreased consumption of petrol and saving annually good money. With the addition of other appropriate renewable technologies and including ever-cheaper batteries, the use of diesel generators will fade out with time. One of the very hot topics for the coming decades will be the supply of drinking water. Desalination today is done using traditional power stations, like the one which I saw recently during a visit in Saudi Arabia where a black cloud of exhaust gas from big diesel generators discolored the sunset. Many GW of clean power will be needed in the future to solve this challenge.

On-grid systems have already demonstrated today that they can produce electricity at the point of use at a lower cost compared to the price which have to be paid otherwise to the utility. This has been described as "grid parity" and is of course not a single number but very dependent on local illumination, finance parameters, and electricity price conditions. The ultimate goal for a 100 % renewable energy supply will be reached in a number of steps where each one is characterized by full economic advantage without feed-in tariff or subsidy:

Supermarkets and cooling stores have an intrinsic advantage for the use of PV, as their electricity need is highest when the sun is shining most. A similar situation will develop for *office buildings, hotels*, and other *commercial applications.* A rooftop system with several 100 kW can, although not supplying all needed annual electricity, lower the load curve especially in times where otherwise a high electricity price would have to be paid (in many places worldwide the price of electricity is increased whenever a certain threshold is overrun). This application can therefore use ~100 % of their self-produced electricity, and the management can compare directly their LCoE—which in Germany is in the range of 13 €ct/kWh—with the price they would have to pay to the municipality, which in Germany for those enterprises is in the range of 15 €ct/kWh. Obviously this can provide a reasonable business case to install a PV system already today—if in Germany the minister for economy, Mr. Gabriel, would not have introduced a "punishment." For systems larger than 10 kW, the user has to pay 30 % in 2015 (40 % in 2017 and later) of the actual feed-in tariff, which in 2015 was 6.17 €ct/kWh. Assuming ~6 €ct/kWh in 2017 the PV owner has to pay ~2.4 €ct/kWh for each self-consumed kWh in 2017 and thereafter. As a consequence, the business case vanishes, and this is the reason why in Germany many stores do not invest any more today in PV. Why those applications do not get even an incentive remains a mystery. In other countries with even more sunshine than Germany, like Spain, there are also regulatory hurdles for self-consumed PV electricity. There is definitely the need to reach in Europe a harmonized legal framework to support self-consumed electricity, which would—not only in Europe—strongly increase the PV installations.

Rooftop PV on single and multifamily houses did pay off only in the past, when an appropriate feed-in tariff was paid despite the fact that LCoE for PV was already lower than the electricity price. The reason for this is that during the year, only ~ 1/4 of the produced PV electricity can be used directly in the house. The other 3/4 are created when no load in the house could use them and have a value which is lower than the LCoE. With strict demand-side management (DSM)—dishwasher, washing machine, and other appliances to be operated when the sun is shining—one can increase the self-consumed electricity by another absolute ~10 %. With the horizon of cheaper batteries and appropriate power electronics, it becomes now possible to increase the self-consumed electricity toward ~70 %, which can be further increased if an electric car is integrated. Once the LCoE for PV electricity together with the LCoS for battery storage is below the price of electricity from the grid—which is the already mentioned second Holy Grail—there will be a strong market increase at a global level.

An interesting development may appear for *municipalities* in the years to come which could nicely end up in new business models for them. As discussed before there will be an excess of PV energy produced even with kWh batteries installed in houses. This electricity could be collected at the low-voltage (380 V) level and with first priority used to power the actual load in their own region and, with second priority, the rest stored in large and cost-efficient MWh batteries for later use. In larger energy regions also, the "power-to-gas" technology can be implemented which would extend the storage time from days to several weeks. This technology uses with third priority excess renewable electricity to produce hydrogen by water electrolysis and combining with CO_2 to obtain CH_4 which can be fed in large quantities to the existing natural gas grid. In a recent study, conducted from IdE (Institute for decentralized Energy [11]), it was shown that in Germany, ~150 energy regions were identified, which have on a local level decided on a timely plan to reach 100 % renewable energy autonomy. It is amazing to see that the population in these 150 energy regions are approaching ~25 million (from the ~80 million) German people. An important aspect in the study was the finding that some of these energy regions consisted of a combination of an urban area together with adjacent rural areas combined together. The former needs a lot of energy and has a lot of capital but lacks the space for enough PV (and wind). In contrast the latter has lots of space, even for smaller wind parks, rooftop, and greenfield PV systems, but has not the capital power. If a municipality, responsible for such an energy region, would then only add a matched CHP plant capacity, there would be the possibility to run this region 100 % autonomously.

There is one remaining challenge which needs to be solved when considering a 100 % supply of renewable energy for all energy needs within an industrialized country: how to power the needs for large *energy-intensive industries*. It has to be noted that already today there is a clear tendency to allocate industries with large needs of electricity near to places with cheap electricity like large hydro stations. Examples are the production of metallurgical silicon in Scandinavia and aluminum from bauxite in Iceland. In the coming decades, there is the very positive development to produce electricity in large quantities in many more regions compared to today. In places like Northern Europe, there will be many GW-offshore wind farms, and in Southern regions we will have GW plants using PV, CPV, and CSP. Large industries will settle ideally in the vicinity of these future large centralized renewable power stations, and many countries which have lost big industries in the past may have the chance to reindustrialize again.

Combining the various applications for the future market development, it is interesting to make a projection for the future market development in order to judge which future prices based on the PEC from Fig. 4.8 can be expected. As described in the preceding market development, there is ample room for a steady and continuous market development. We will use for demonstration purposes the following simplified growth model as shown in Table 4.3.

The extrapolated PEC in Fig. 4.8 tells us that we can expect the following future fair prices approximately in the years (in times of overcapacity we may see the given prices even earlier):

- 0.6 $/W in ~2020 at ~600 GW(cum)
- 0.5 $/W in ~2022 at ~1000 GW(cum)
- 0.4 $/W in ~2027 at ~2000 GW(cum)
- 0.3 $/W in ~2031 at ~4000 GW(cum)

Table 4.3 Simplified future growth for PV until 2050

Year	Annual market [GW]	Cumulative global market [GW]	Average annual growth in preceding decade [%]
2010	17	43	>50
2020	100	670	20
2030	460	3400	16
2040	1000	10,600	8
2050	1400	23,000	4

Using our simplified LCoE model, we can expect a further bisect from today's numbers in the late 2020s/early 2030s toward ~4-8 $ct/kWh in northern and ~2-4 $ct/W in southern regions which makes PV the least costly electricity generation technology. This finding has also been concluded by IEA in their hiRen scenario [12], but at a later time in the 2050s which is no surprise as they are only assuming a much smaller growth toward ~4.6 TW in 2050 with a corresponding higher module and system price in 2030.

4.4.3.2 Future PV Technology Development

In order to further follow the PEC for PV modules shown in Fig. 4.8, there is the continuous need to drive technology to lower and lowercost levels. One important lever for many applications is the efficiency of the PV cell as this has a pronounced effect on all area-related cost additions, like module making, BOS cost, and installation. We will discuss the efficiency development starting from current status (2014) toward 2020 and 2030. Each of the four major classes will be described with three different technologies:

- High-efficiency silicon
  - Heterojunction cells (HIT)
  - Interdigitated back contact (IBC) and HIT back contact cells
  - Future silicon wafer plus tandem thin-film cells
- Standard silicon
  - Standard multicrystalline cells
  - High-performance multicrystalline cells
  - PERC (passivated emitter rear contact) cells
- Thin film
  - CIGS (copper indium gallium selenide)
  - CdTe/CdS
  - Thin-film tandem

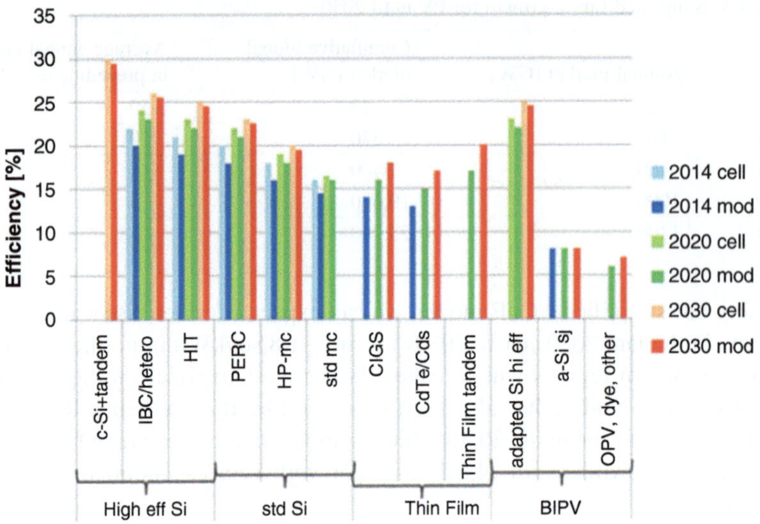

Fig. 4.9 Efficiency development for the various PV cell and module technologies from 2014 to 2020 toward 2030

- Building-integrated products
  - Adapted silicon cells (and also adapted thin film)
  - Amorphous silicon
  - OPV, dye, and others

For the c-Si wafer-based technologies, the respective light color in Fig. 4.9 shows the cell efficiency, and the dark color shows for all technologies the module efficiency. The following points should be highlighted:

1. The standard multicrystalline silicon will stop in the 2020s and continue with high-performance structures.
2. The c-Si plus tandem has just started to be investigated in laboratories, for example, the combination of c-Si with perovskite. In principle it is possible to reach efficiencies beyond 30 % with such structures.
3. Today and also in 2020, the cell efficiencies are significantly higher compared with the total area module efficiencies. This is due to the non-active areas, e.g., between the cells and around the circumference of the module which cover typically about 7–8 % of today's 60 cell modules. In addition the so-called cell-to-module power ratio (defined as (module power)/(cell power × number of cells)) is today in the range of ~98 %. With proper light management technologies, it will be possible to redirect light from these non-active areas on to the solar cells thereby increasing cell-to-module power ratio toward 103 % (ITRPV [10]) and ultimately the module efficiency up to levels near to the cell efficiency.
4. Thin-film tandems have been manufactured in the past using amorphous silicon together with microcrystalline silicon films. This combination could so far not

reach efficiencies which are obtained with single-junction CIGS and CdTe/CdS modules and has been ended. However, in the future other combinations with different bandgap materials will offer the possibility to reach module efficiencies in the 20+ % range.

5. The portfolio of products for the BIPV (building-integrated PV) application is shown separately and consists of similar product components as discussed before (like crystalline silicon solar cells and thin-film raw modules), integrated in façade or overhead structures or specifically designed PV devices (like amorphous silicon, dye, and organic solar cells).

The c-Si solar cells are well suited for cold-wall facades with good ventilation to keep the temperature at low levels. Integration into insulating façade elements requires a number of specific needs with the most important ones:

- Various degrees of semitransparency
- Homogeneous appearance
- Different colors
- Low-temperature coefficient for efficiency
- Different sizes

It is obvious that c-Si solar cells are not the ideal product to fulfill the specified needs as described above. Thin-film modules and even more so new technologies like dye solar cells (DSC) and organic PV (OPV) open up many more possibilities to integrate PV into the building envelope. DSC and OPV can be produced in the future using roll-to-roll printing technologies using flexible substrates.

One thin-film technology, single-junction amorphous silicon (sj-a-Si) on glass, has often been declared as "dead." However, utilizing its characteristics properly, there is still a chance to use this technology, for the following reasons:

- Sj-a-Si has a very thin layer thickness of less than half a μm and can therefore easily (e.g., lift-off technologies) be made semitransparent from a few up to 50 %.
- This material has a homogeneous appearance.
- Different colors cannot be produced with the absorber material (like DSC and OPV) but by using colored substrate glasses.
- The material has one of the lowest temperature coefficients (~0.2 %/°C).
- It was already demonstrated with the Applied Materials "SunFab" that modules can be produced on glass substrates with 5.7 m^2: this allows the façade-producing company to cut from large areas the needed pieces for completing the final double or triple insulating glass element.
- Very interestingly the cost and price per m^2 are the lowest for all materials known so far: with a cost per W at least similar (normally lower) compared with other PV products with higher efficiency, there is a significant lower cost per area possible, e.g., compared to c-Si modules, a cost decrease of more than 50 %.

It remains to be seen whether dedicated companies will utilize these specific advantages for sj-a-Si to be integrated into BIPV products.

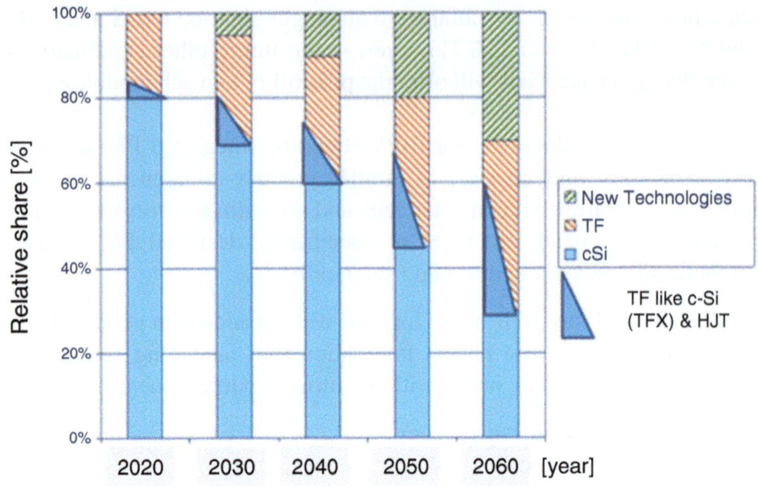

Fig. 4.10 Relative market share for various cell and module technologies

An interesting question has often been discussed which deals with the split in technology for the future. I have been following over my 35-year experience in the PV industry quite contradicting forecasts: when in the 1980s, the advent of thin-film technology started and reached quickly ~30 % relative market share, there were many experts predicting a take-over by this technology in only a short time. However, when in the first decade of the new century, the market exploded, there was the need of quick capacity ramp-up, which was only possible with a mature technology. This was the main reason why c-Si technology quickly gained 90 % relative market share. In these days I hear a group of experts arguing that c-Si will even increase this relative market share. Another group of experts claim that the so-called third generation of PV—OPV (organic PV) and DSC (dye solar cells)—will be able to take the biggest relative market share in a short period of time. I personally did and will not follow these experts at any time and argue that it will be the specific customer need which will drive well-adapted technologies. This is why I see with the differentiation of applications the need for different technologies. Figure 4.10 shows in a simplified picture my view for c-Si, thin-film, and new technologies like OPV and DSC but also including CPV (concentrated PV). There will be no new generation—which would imply that the preceding generation will pass away—but a balanced contribution of the various technologies. There will be also an increasing challenge to clearly distinguish c-Si and thin-film technology as there will be interesting combinations from the two as described before (e.g., c-Si heterojunction or tandem cells).

The important message from Fig. 4.10 is that:

- Although the relative share of c-Si will shrink, there will still be a sound growth given the overall growth of the PV sector.
- Thin film will gain market share due to the growth of BIPV and other applications using flexible substrates.

Table 4.4 Cost limit for the value added steps for c-Si wafer-based modules with back-contacted cells

Value added	description	[$ct/W]
Wafer	200 × 200 mm^2, 80 μm thickness/70 μm kerf loss, 12 $/kg polysilicon, module efficiency 25 % → silicon material cost	2
	Other material for wafer making	2
Cell add-on	No silver, but Cu or Al for metalization	2
Module add-on	Double glass a 1.4 mm	4
	Encapsulation, etc.	3
	Cable, back rail, etc.	3
Total OPEX	(Operational expenditure)	16

- c-Si and thin film will combine in a number of future devices.
- New (but already existing) technologies like OPV, DSC, and CPV will gain market share in coming decades.

Looking to the price experience curve from Fig. 4.8, there is the interesting question of how long the prices will be able to follow this straight line. The following estimation in Table 4.4 should help to better understand the continuing price decrease at least down to levels of ~20+ $ct/W. From today's point of view, we can assume the following cost estimate for a future back-contacted solar cell:

If we assume for a mass-produced product like PV modules a share of 80 % from total cost for operational expenditure (OPEX), we have for the rest another 4 $ct/W which makes a total cost of ~20 $ct/W for c-Si modules. The comparison with the price of 20+ $ct/W which is reached at a cumulated volume of ~20 TW gives a first qualitative indication that there is room for some margin equated by the "+". This appraisal is without tandem structures which could further decrease this number if the additional efficiency adds less to the cell add-on compared to the saving in the module add-on (and also the BOS). Similar exercises can be made also for the other technologies.

If we allow the split in 2050 according to Fig. 4.10 and assume only ~50 % for c-Si and TF-like c-Si, the margin would be even larger if the industry were capable to reach already at that volume the abovementioned lowest cost structure. The assessment for prices, cost, and volume for 2050 should only be seen as a proof of concept and a qualitative understanding for the further price development based on realistic assumptions for growth rate and competitiveness of PV compared to other renewable technologies—noteworthy to say that traditional technologies like clean fossil and new nuclear are hopelessly too expensive already in the late 2020s/early 2030s.

One important consequence from this most probable development is the fact that new investments for clean fossil and new nuclear which are typically based on depreciation times of 30+ years (like the 35-year agreement for the new nuclear power stations at Hinkley Point in the UK) do no longer pay off as they will be closed down before the end of their depreciation time due to economic inefficiency. Once the financial world will have understood these simple facts, it will be very hard to find in the future money for investing in these traditional power stations which will naturally be their end. Looking to the hundreds of billions of dollars which are currently planned to be invested on a global level, it is frightening to see how a good part of this money will evaporate in only a few years from now.

4.5 Realistic Contribution Of 20 % Secondary Energy with PV Within A 100 % Renewably Powered World

4.5.1 *Integration of Increasing Levels of Variable Power from Renewable Energy to the Annual Load Curve of a Country*

The integration of increasing levels of variable renewable energy sources calls for radically new ways for the traditional electricity market. It is no surprise that after many decades of successful electrification in the OECD countries including the emerging markets with a huge hunger for electricity, a well-functioning network emerged. Huge power stations—fossil and nuclear—provided (at least in the past) cheap base load, and the distribution was organized top down: high-voltage transmission over large distances to the medium-voltage distribution network down to the municipality low-voltage grid. Everything was optimized to serve the required load by a combination of the abovementioned base load with medium and peak power stations. But not only the technology was streamlined to these requirements, also the trading of electricity at the stock exchange was well adapted to these boundary conditions. When in Germany in 1991 with the electricity feed-in law (Stromeinspeisegesetz (StrEG)) and later in Europe with a binding directive, the priority for renewable electricity against fossil and nuclear was formulated, there was little concern by utilities and politicians because the produced quantity from wind and solar was pretty small at that time, much less than 1 % of the annual electricity consumption. However, this changed dramatically with the feed-in law (EEG) in Germany and other European countries, which induced an unexpected market growth. This resulted in a number of challenges, exemplified with the example of Germany, which had at the end of 2015 about 29 % renewable (2.9 % hydro, 6 % biomass, 12.3 % wind, 6.0 % PV, and other 1.9%) and 71 % traditional electricity-producing technologies to provide the ~600 TWh annual electricity:

- Although PV and wind provided only 6 and 12.3 % electricity in 2015, both technologies had a cumulative capacity of >40 GW each.
- Knowing that on a typical weekend the German load only needs maximally ~40 GW in the afternoon hours, it is obvious that on a sunny summer weekend day, the total load can be provided—even only for a short period of time—by PV alone.
- With the priority of renewable energy, there are in consequence only two possibilities for traditional power stations: they either have to shut down or

they try to sell their electricity at the stock exchange even at negative prices. As the shutdown would only be for a short period of time, it would be quite costly to ramp down and shortly after to ramp up again a conventional power station. In order to minimize their losses, utilities are willing to not shut down and sell those kWhs even if they have to pay for it. However, there is also an additional very unfortunate development with increasing levels of renewable technologies: curtailment of power from renewable production sites when the grid is not able to transport the electricity away from the production site. For example there was in the year 2014 in Germany a curtailment of ~1,200 GWh for wind, ~250 GWh for PV and even ~100 GWh for biomass.

In the following we will analyze the German case if we increase the annual electricity production toward 100 % renewable technologies and the resulting necessities to cope with variable power sources like PV and wind.

The annual load curve for an industrialized country like Germany is shown schematically in Fig. 4.11. The two extremes are the continuous load of ~35 GW (also called base load and obtained by $t = 8760$ h) and the maximum power of ~90 GW which is only needed for a few hours within a year (also called peak load at $t = 0+$ h). The necessary annual energy which is obtained by integrating this annual load curve is ~600 TWh. The weekly load curve shows higher power needs during winter compared to summer and exhibits during the weekend a daily load curve which is ~40 % reduced compared to the 5 workdays. The daily load curve shows typically a pronounced peak at noontime. The example of Germany, which has by the end of 2014 a total of ~38 GW PV systems installed, shows impressively that on some sunny summer weekend days most of the load can be served by only PV, although the annually produced PV energy of ~35 TWh adds only ~6 % to the

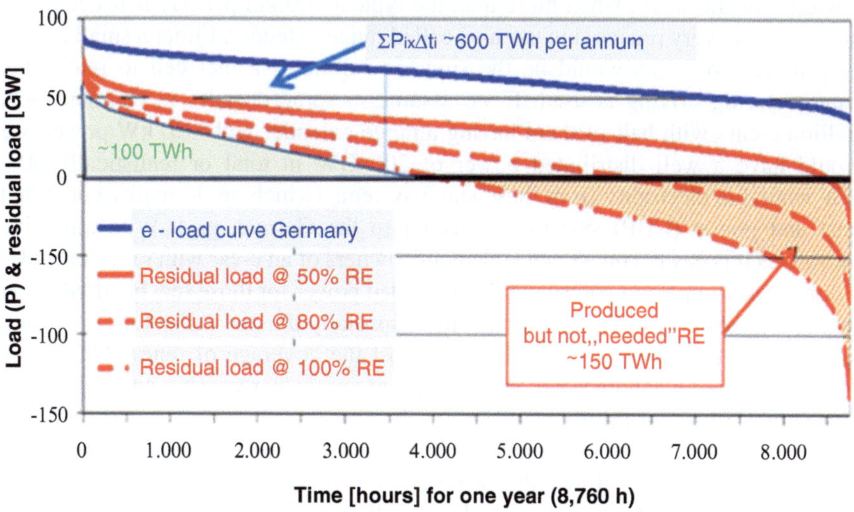

Fig. 4.11 Load curve and residual load for three levels of contributing renewable energies (example Germany, data from [13])

annual electricity needs. By adding more and more PV power, it will occur more frequently that available solar power cannot be used directly from the load.

In order to increase the fraction of annual PV energy toward 1/3, corresponding to 200 TWh, we would need ~220 GW-installed PV systems. In this picture wind would supply most of the other 2/3 of annual electricity with about the same installed power compared to PV. The comparison with the annual load shows that the desirable high-installed power can only be realized with a cost-effective storage. Similar conditions are also found in Italy and other European countries.

In order to supply the future electricity substantially with renewable energy, the variable renewable technologies will increasingly produce electricity, which will not be needed at the time of production. The example in Fig. 4.11 for 100 % RE shows that ~ half of the year, there will be oversupply (negative residual load), while for the other half year, there will be a positive residual load, which cannot be served by the installed renewables in those hours. For the example shown, the excess energy is about 150 TWh (red shaded area), while the needed positive residual load is about 100 TWh (blue shaded area). With the assumption that the average efficiency for all used storage devices (batteries, pumped hydro, compressed-air store, and "power to gas") is approximately 66 %, a complete coverage for the annual load curve with only renewables would be feasible. The example shown in Fig. 4.11 should only be understood as a schematic picture. For the envisaged smart grid in combination with the future smart home (utilizing weather forecast plus DSM (demand-side management)), there will be good possibilities to further optimize the needed storage capacity.

One possibility to minimize the needed electricity storage, especially in times with no sun and no wind for a couple of days, is the integration of future range extenders in electric vehicles. As described earlier, future transportation will increasingly be powered by e-cars. In order to increase the range of the car for few days of the week, when more than the typical 100 km per day is needed, it is already today very practical to use a so-called range extender. Either a small engine as generator or—this would be the superior solution—a fuel cell to reload the battery during driving is used. If we assume in some decades in Germany ~20 million e-cars with half of them having a range extender with ~20 kW power, we would have a well distributed power of 200 GW in total or realistically 50+ GW available electricity from these small systems (which are in reality combined heat and power (CHP) systems) to feed into the German grid. With attractive incentives it is easily conceivable that many owners of an e-car with range extender would even compete to contribute to the needed power for those few hundred hours of a year. It should also be highlighted that especially the fuel cell solution—sitting overnight in the garage—would help to boost the coverage of renewable energy sources in individual homes not only in terms of energy per year but on a daily basis for the needed power locally.

In northern countries with more wind and less sun compared to Germany, one would adjust the ratio of installed wind and PV power accordingly. The Sunbelt region, where most of the global population is living, would use more PV compared to wind.

4.5.2 Price Experience Curve for New Battery Technologies: Evidence for Cost-Effective Electricity Storage

The global battery market from 1990 to 2013 together with a projection for 2015 and 2020 is seen in Fig. 4.12. While in 1990 the lead-acid battery (LAB) dominated the market, the further development showed an over-proportional growth for Li-ion batteries (LIB). All other technologies like NiCd, NiMH, and others contributed only minimally to the global market.

In Figs. 4.13 and 4.14, the annual battery market between 2000 and 2013 for only Li-ion batteries is shown for total capacity (in MWh) and turnover (in million dollars), respectively

Two special battery types have been analyzed in more detail: first the Wh batteries for cellular phones (blue bar in left subimages) and second the kWh batteries for automotive applications (red bar in right subimages). The first type was chosen because this battery type has already a good track record from 2000 to 2013, while the kWh battery is the one particularly needed for e-mobility and electricity storage in houses to increase self-consumption of variable renewable electricity sources, e.g., PV. From the annual turnover and the respective volume in MWh for a particular year, the average price in $/kWh for this year can be calculated.

With the data shown in Figs. 4.13 and 4.14, the $/kWh numbers can be associated with the corresponding cumulated volume. If these data points are

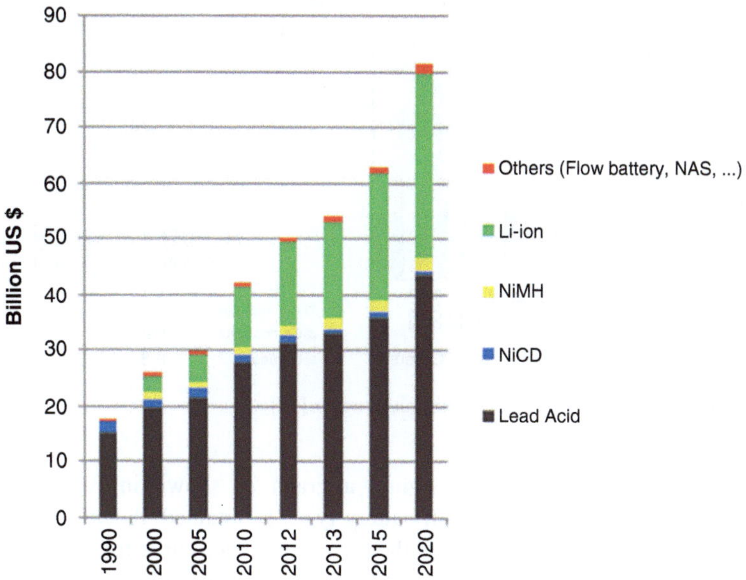

Fig. 4.12 Annual battery market in bn$ differentiated by technology during 1990 and 2013 with a projection to 2015 and 2020 [14]

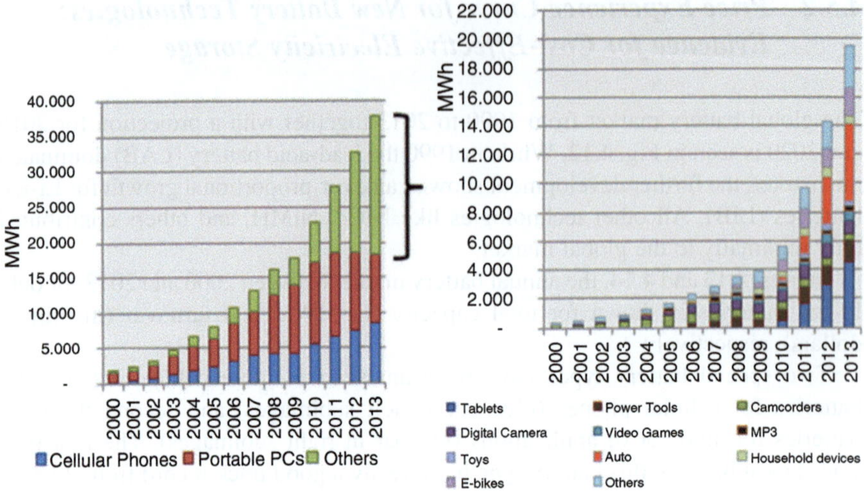

Fig. 4.13 Annual battery market in MWh for LIB during 2000 and 2013; (*left*) all applications (*lower blue bar* Wh-cells for cellular phones); (*right*) differentiating the *left green bar* "others" (*red bar* for kWh cells for automotive) [14]

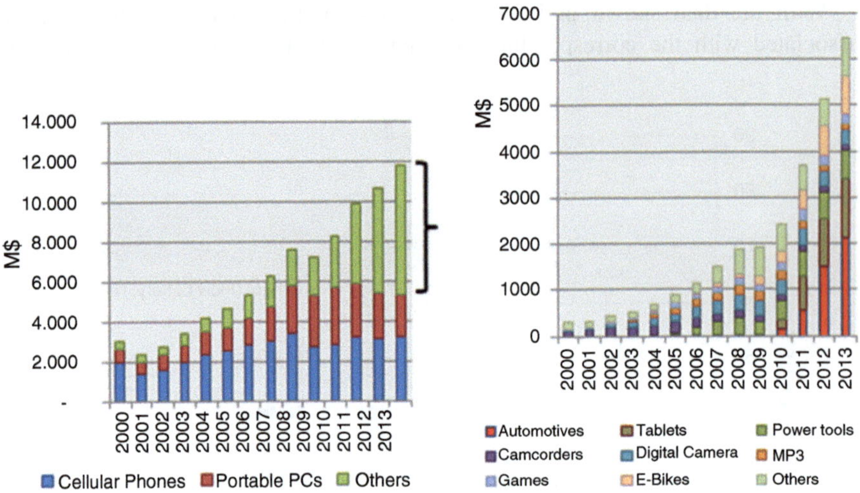

Fig. 4.14 Same as in Fig. 4.13 but in million turnover [14]

plotted in a double logarithmic figure, a graph as shown in Fig. 4.15a (for Wh-battery type) and 15b (for kWh-battery type) is obtained. A least square fit to the data points reveals a straight line, the price experience curve for the two battery types. From the slope the price experience factor can be calculated which is ~25 % for Wh batteries and ~15 % for kWh batteries.

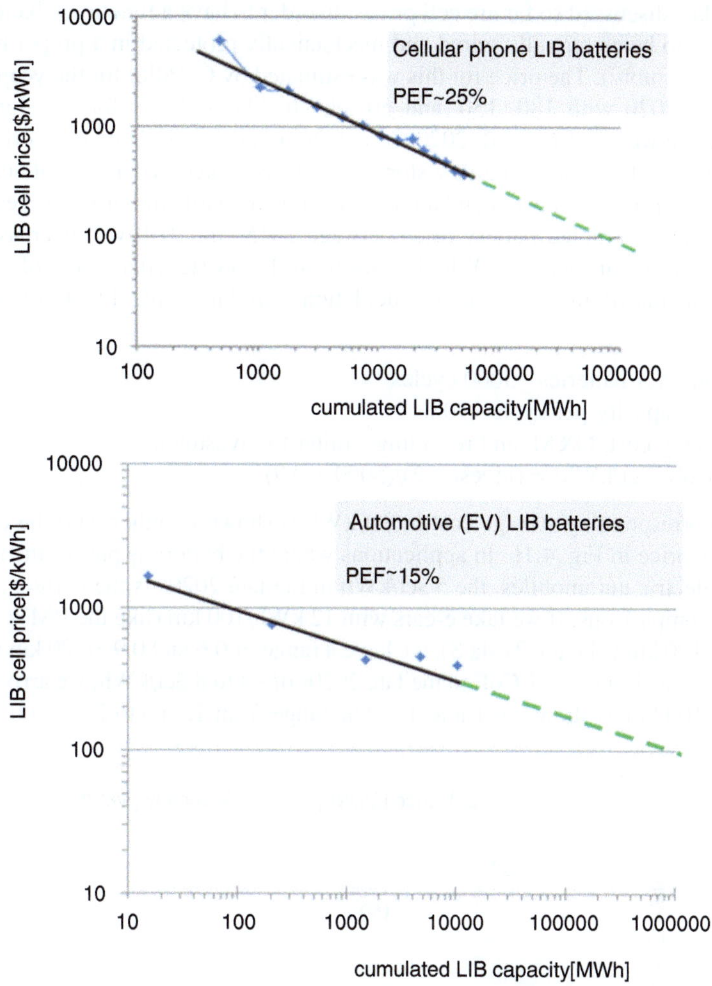

Fig. 4.15 Price experience curve for two different Li-ion batteries: (**a**) above for Wh batteries for cellular phones and (**b**) below for kWh batteries for automotive applications (data from Avicenne, Christophe Pillot [14])

In analogy to other high-tech products, it will be highly probable that the future prices will be around the extrapolated PEC. In both cases the extrapolation gives a battery price of 100$/kWh at a cumulated volume of ~ 1 TWh (1 million MWh). The question arises which annual growth rate will be needed to reach from a cumulated volume in 2013 (60 GWh for Wh batteries and ~10 GWh for kWh batteries) in both cases the 1TWh-level. If we—arbitrarily—want to know this growth rate by the year 2030, a growth rate for the cumulated volume of 18 and 31 % per year is needed for the Wh- and kWh-battery type, respectively. This growth is ambitious (especially for the kWh battery), but in comparison to PV modules, still moderate: the cumulated PV market had an average annual growth rate of 41 % between the years 2000 and 2010.

The data discussed so far are cell prices. In order to have a functional battery, the cells have to be electrically wired and mechanically protected in a proper housing (battery assembly). The price for this was estimated by C. Pillot for the years 2012, 2015, and 2020 with 180, 150, and 50 \$/kWh. Major driver for the significant decrease between 2015 and 2020 is a substantial growth, mainly driven by e-mobility (and stationary battery storage) and associated "economy of scale." If we keep the price ratio between battery assembly and cell similar to the year 2020 (1 to 4), the assembled battery price between 2015 and 2030 will decrease from ~400 \$/kWh toward ~125 \$/kWh. The resulting LCoS (levelized cost of storage) can be calculated in a simplified calculation (similar to the LCoE for PV) as follows:

- Lifetime for batteries ~5000 cycles
- Usable capacity per cycle ~90 %
- Financing cost, O&M, and recycling similar to investment
- $\rightarrow$ LCoS [\$ct/kWh] = (invest × 2)/(5000 × 0.9)

The corresponding storage cost in \$ct/kWh is shown together with the cell and assembly price in Fig. 4.16. In applications where the battery is part of the system, e.g., in electric automobiles, the 5 \$ct/kWh in the late 2020s is the correct number for cost comparisons. If we take e-cars with 12 kWh/100 km (like the BMW i3) and 18 kWh/100 km (like the Tesla S), we have a range of 0.6 and 0.9 \$/100 km storage cost. By considering PV LCoE in the late 2020s of ~4 to 8 \$ct/kWh, we arrive at the cost per 100 km as shown in Table 4.5. The range from 1.1 up to 2.3 \$/100 km can

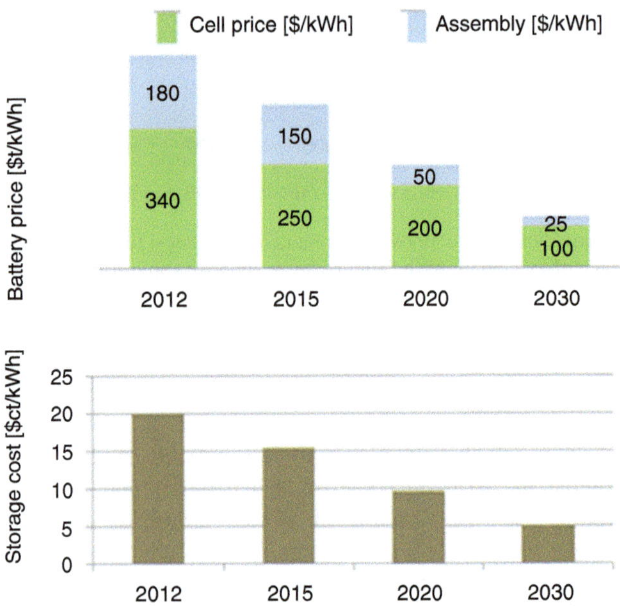

Fig. 4.16 Battery price [14] and LCoS for kWh-LIB until 2030

Table 4.5 Storage cost for e-cars (small like BMW i3 and big like Tesla S) plus PV cost in Southern and Northern countries per 100 km driving distance

Type	Need [kWh/ 100 km]	LCoS [$/ 100 km]	PV south LCoE [$ct/kWh]	PV south LCoE [$/ 100 km]	South total cost in $ per 100 km	PV north LCoE [$ct/kWh]	PV north LCoE [$/100 km]	North total cost in $ per 100 km
BMW i3	12	0.6	4	0.5	1.1	8	1.0	1.6
Tesla S	18	0.9	4	0.7	1.6	8	1.4	2.3

LCoS levelized cost of storage; *LCoE* levelized cost of (PV) electricity be compared to the cost of a petrol-driven car. We should for comparison reasons not take the price for petrol including the tax but only the fuel cost without, which will be around 0.5 $/l. Considering a range between 5 and 8 l/100 km, we have 2.5 and 4 $/100 km for small- and bigger-sized cars. It is clearly seen that the fuel cost for comparable e-cars is significantly better. As we have included the battery investment in the fuel cost (LCoS), the price of the e-car without battery will be in the future lower compared to an equally powered petrol-driven car. Future electric wheel hub motors will even widen the gap. In total there will be no future for fossil-powered automobiles just and because they will be more expensive.

It should be noted that the total LCoS can increase up to a factor of 2, for example, in small-scale stationary home battery systems. This is because we need additional electronic power devices to communicate with the PV DC/AC inverter and the installation of small-scale systems add as always over proportionally to the specific price of the total system.

4.5.3 The 100 % Renewably Powered World

In 4.1.3 we concluded for the future needs for ten billion people a secondary energy of 150 PWh. In Fig. 4.4 and Table 4.1, we have summarized the renewable offerings in terms of sustainable supply. The resulting 3,100 PWh are fortunately over 20 times more compared to the needs. In order to allow for regional optimization, it is advisable to include besides the dominating solar and wind also other renewables even if they will not reach the LCoE numbers compared to PV in the future. In a recent book by W. Hoffmann [15], a more detailed discussion was elaborated which resulted in a pragmatic split as follows:

- **20 + % solar PV for decentralized electricity**
- **20-% solar concentrating large scale (CPV, CSP) for energy-intensive industries (electricity and process heat)**
- **20 % solar thermal for domestic heating and cooling including medium process heat**
- **20 % wind on- and offshore**
- **20 % all other renewables like hydro, biomass, geothermal, and marine tidal and wave**

For the 20 % solar PV we have already given in Table 4.3 a simplified growth model how to reach in 2050 the needed 23 TW cumulated PV installations to provide the 30 PWh annual electricity (conservatively 1.3 $kWh/W_{PV\ installed}$ as a global average is used). In Fig. 4.17 this growth is shown, and it is obvious that a leveling of 23 TW cannot realistically be described with this growth curve. If for simplicity reasons we assume a 40-year replacement period for PV modules, we would have to start in 2051 with the same small annual addition as in 2010, a reduction from 2050 with ~1400 GW down to ~20 GW. The same procedure would happen again in 2091. This is obviously not a realistic picture and can only serve to

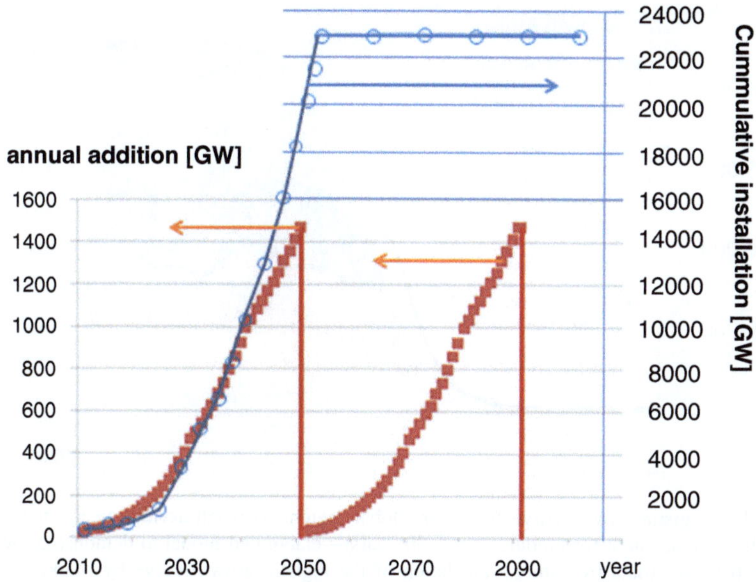

Fig. 4.17 Annual growth (*left*) with average growth numbers per decade (as in Table 4.3) to reach 23 TW_{cum} (*right*)—quick and dirty exercise to demonstrate realistic feasibility

Table 4.6 Parameters to calculate the respective logistic growth curve for each of the region(s)

Region(s)	Cumulative PV installations in TW ("G") in 2050+	Growth slope ("k")	Time constant ("c")
Africa	8.6	0.22	2050
ROW	4.9	0.28	2035
Americas	1.2	0.20	2035
Latin America	1.7	0.17	2040
PRC	2.0	0.22	2030
India	3.1	0.25	2035
EU	1.5	0.15	2032
	23		

quickly demonstrate that the industry up to 2050 could cope with the implementation of the 23 TW PV systems.

A much better and more realistic growth curve can be modeled by using a so-called logistic growth curve. In addition it is possible for different regions to use different parameters for the growth and asymptotically reached maximum cumulative installed volume. If we equate the respective cumulative PV installations after 2050 with data for population from the UN [4], we arrive at cumulative PV volumes as given in the first column from Table 4.6. Here the assumption is made that the split for the 23 TW cumulative PV installations is proportional to the split in population in the respective regions. A description of the model is given in

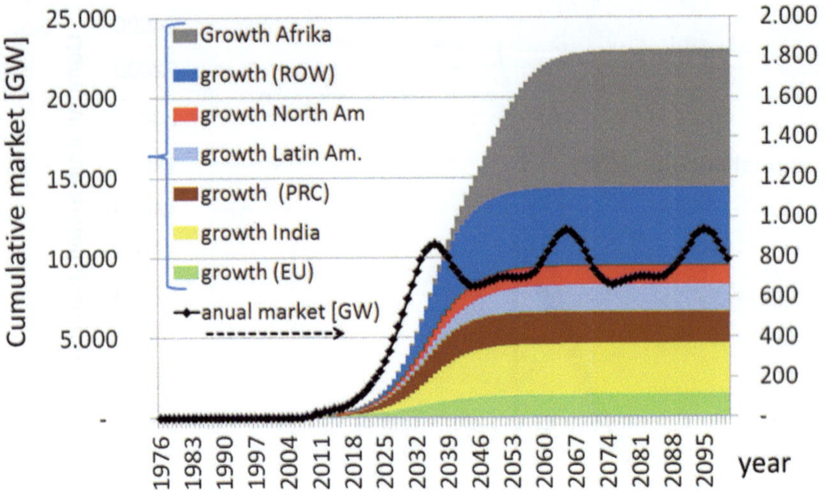

Fig. 4.18 Logistic growth curve for seven global regions to reach asymptotically the 23 TW (including replacement for modules every 30 years)—elaborated model to demonstrate realistic growth behavior (data by author, calculation of the logistic growth curve by Markus Fischer, Hanwha Q Cells GmbH)

ITRPV [10], where the following equation is used for calculating the logistic growth curves for each of the respective region(s):

$$N(t) = {}^{G}\Big/_{\left(1+\ e^{k(c-t)}\ \right)}$$

Together with an assumed module replacement time of 30 years, the annual total market together with the cumulated PV installations can be calculated and is shown in Fig. 4.18.

The following conclusions can be drawn:

- No "sharp" reaching of 23 TW in 2050, but a more asymptotic culmination.
- The annual global market (right ordinate) cycles around 800 ± 150 GW.
- The chosen set of parameters is done in a pragmatic way using reasonable assumptions for the various regions.

As a summary the future picture for supplying the secondary energy with only renewables for the ten billion people looks quite different as shown in Fig. 4.19. In particular the share of ~2/3 electricity for the secondary energy is important to note which is easily understood from the fact that the important renewable technologies like PV and wind produce the convenient secondary energy electricity. From a consumption point of view, it is also important that as described earlier we exchanged today's one quarter of secondary energy for mobility from oil to electricity (or electricity-based fuel). The rest, which is heat and process heat, is provided by the other renewable technologies.

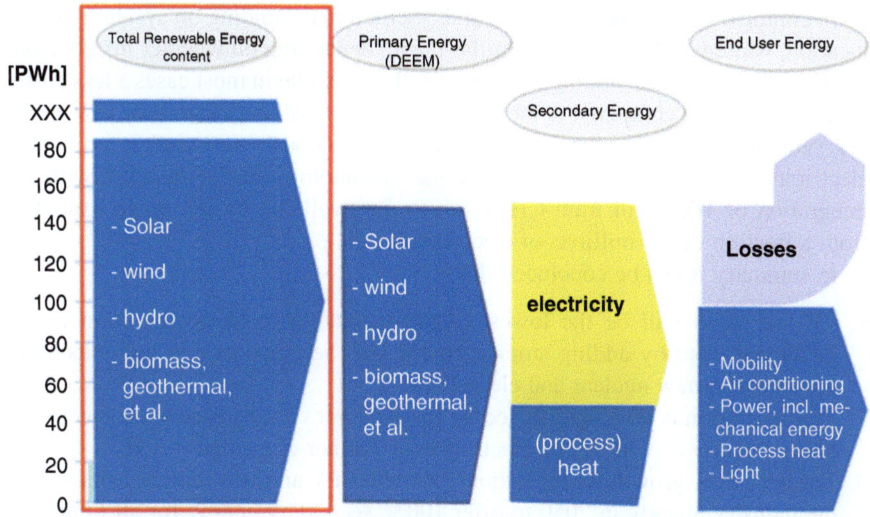

Fig. 4.19 Primary, secondary, and end energy in a future 100 % renewably powered world

	Technology	LCOE in today's currency [\$ct/kWh]
Traditional	Clean coal with CSS	> 10
	Nuclear fission	> 10
Photovoltaics	Southern areas (~2 kWh/W_{PV})	2 – 4
	Northern areas (~1 kWh/W_{PV})	4 – 8
Wind	On-shore (~2 kWh/W_{wind})	3 – 4
	Off-shore (~4 kWh/W_{wind})	5 – 8
Storage	Small (~kWh+)	> 5
	Large (~MWh)	< 5

Fig. 4.20 Future LCoE for traditional sources, PV and wind, together with electricity storage cost

Figure 4.20 highlights at a glance where the electricity-producing industry will develop the various LCoEs for the different technologies in the late 2020s—only 15 years from now. The very good news is that it will be the economics which will make renewables superior to nuclear and clean fossil. As already discussed earlier, the LCoE for nuclear and clean fossil will be well above 10 \$ct/kWh—with the cost for the high- and medium-voltage transmission lines even to be added.

In contrast, PV will be between 2 and 8 $ct/kWh, depending on system size and location, while wind will be in a similar range, depending on whether installed on- or offshore. Even when adding storage cost, there will be in most cases a less costly kWh of electricity compared to the conventional technologies. For those few weeks per year in a country where no sun and simultaneously no wind is producing electricity, there will be the "power-to-gas" technology in combination with the integration of 1000 s of mini-CHPs, which are available by the range extenders from a fraction of the millions of electric vehicles.

In summary it can be concluded that:

1. PV and wind will be the lowest LCoE-generating technologies latest in the 2020s and also by adding storage cost it will be considerably less expensive compared to new nuclear and clean fossil.
2. Transportation is no longer fueled by fossil sources but by renewable electricity either directly or with H_2 by electrolysis or further converted to CH_4.
3. The necessary growth for the major renewables PV and wind can be achieved by the industry already in 2050 to offer 100 % secondary energy for all future ten billion people and the global industry.
4. The investment for 100 % renewable supply (~23 TW PV, ~12 TW wind power, ~15 TW CPV/CSP, etc.) will be enormous; however, according to N. Stern [16], the cost for repairing the damage due to climate change if we continue with fossil and nuclear as today will be far higher.
5. Once the financial world will have understood these facts, there will be no longer investment money for traditional energy technologies like fossil and nuclear but only in renewables—this will be enhanced by the fact that unlike fossil and nuclear technologies, renewables do not add to the climate and environmental problems.
6. Renewables are by nature available everywhere in sufficient quantities—they add therefore significantly to the security of supply in all countries.

Acknowledgment The author would like to thank Christophe Pillot from Avicenne (Paris) for providing the battery data for Figs. 4.12–4.16 and Markus Fischer from Hanwha Q Cells for kindly calculating the logistic growth curve in Fig. 4.18. Figure 4.8 is with permission of ITRPV (Stefan Raithel). Figures 4.1–4.4, 4.6, 4.19, and 4.20 are with permission of Scrivener Publishing (Wiley) [15].

References

1. IEA (International Energy Outlook), World Energy Outlook (2011)
2. E.U. von Weizsäcker, A.B. Lovins, L.H. Lovins, *Factor four: Doubling wealth – halving resource use: The new report to the Club of Rome* (Earthscan, London, 1998)
3. E.U. von Weizsäcker, K. Hargroves, M. Smith, *Factor five: Transforming the global economy through 80% improvements in resource productivity* (Earthscan, London, 2009)
4. United Nations, Department of Economic and Social Affairs, Population Division, *World Population Prospects: The 2012 Revision, Highlights and Advance Tables.* Working Paper No. ESA/WP.228 (2013)

5. BP Energy Outlook 2035 (2015) www.bp.com/content/dam/bp/pdf/Energy-economics/energy-outlook-2015
6. Advisory Board to the German Government on Global Change (WBGU), Flagship Report 2011, *World in Transition: A social Contract for Sustainability,* ISBN 978-3-936 191-37-0, http://www.wbgu.de/fileadmin/templates/dateien/veroeffentlichungen/hauptgutachten/jg2011/wbgu_jg2011_en.pdf
7. D.M. Chapin, C.S. Fuller, G.L. Pearson, *A new silicon p-n junction photocell for converting solar radiation into electrical power* (Bell Telephone Laboratories, Inc., Murray Hill, 1954)
8. EPIA/SolarPower Europe, Global market outlook for photovoltaics 2014–2018. ISBN 9789 0822 28403
9. W. Hoffmann, S. Wieder, T. Pellkofer, Differentiated price experience curves as evaluation tool for judging the further development of crystalline and thin film PV solar electricity products, 24th PVSEC Hamburg (2009)
10. International Technology Roadmap for Photovoltaic (ITRPV) (2015), http://www.itrpv.net
11. Institut dezentrale Energietechnologien (IdE) (2012), www.ide-kassel.de/projekte/projekt/100-erneuerbare-energie-regionen/
12. IEA, *Energy Technology Perspective,* (2014)
13. Niklas Hartmann, Dissertation Universität Stuttgart (2013)
14. Personal communication Christophe Pillot, avicenne, (2014)
15. Winfried Hoffmann, The economic competitiveness of renewable energy—pathways to 100% global coverage, Wiley/Scrivener Publishing (2014), ISBN 978-1-118-23790-8
16. Nicholas Stern, Stern review on the economics of climate change (2006), http://www.hm-treasury.gov.uk

Chapter 5
Designs of Various Hybrid Photovoltaic-Thermal (PV/T) Solar Collectors

Mohd. Yusof Hj. Othman and Faridah Hussain

5.1 Introduction

One of the crucial factors that affect the popularity of photovoltaic module is its relatively low efficiency. Until today, commercially available photovoltaic module efficiency claimed by manufacturers is from 6 to 16 %. However, the claimed efficiency is at a temperature of 25 °C. In reality, especially for countries with hot weather, their ambient temperature during a sunny day would be more than 35 °C. The rising of the PV temperature will result in the drop of module efficiency. By calculation, it was concluded that, for typical silicon-based PV, the efficiency of the cell would decrease 0.5 % with 1 °C increase in temperature [1].

Nowadays, most of the commercial PV modules are made up of a thin layer of silicon semiconductor. The bandgap of a semiconductor tends to decrease with the increase of temperature. When the temperature increases, the amplitude of atomic vibration increases; therefore, the space between atoms will also increase. These will reduce the size of the bandgap. When the bandgap gets smaller, most of the electrons from the conduction band will fall back to the valence band and recombine with holes. These reduced numbers of moving electrons in the conduction band hence reduce the efficiency of the solar cells.

In order to minimize PV module from losing its efficiency, a simultaneous cooling system using air or water as the heat transfer liquid can be implemented. The heat output from the system can be collected and stored as thermal energy. This

M.Y.H. Othman (✉)
Solar Energy Research Institute (SERI), Universiti Kebangsaan Malaysia, 43600 Bangi, Selangor, Malaysia
e-mail: myho@ukm.edu.my

F. Hussain
National Metrology Laboratory, SIRIM Berhad, 43900 Sepang, Selangor, Malaysia
e-mail: faridahh@sirim.my

A. Sayigh (ed.), *Photovoltaics for Sustainable Electricity and Buildings*,
DOI 10.1007/978-3-319-39280-6_5

advanced system is known as PV/T technology. The concept of PV/T system had been studied, discussed, and published for almost 4 decades now. Review papers related to research's work on PV/T collectors had been written by [2–5]. Both water and air are suitable to be used as the cooling fluid. However, air-based PV cooling system is simpler and economical due to its minimal usage of material and low operating cost. A high-thermal-conductivity material together with a natural or forced flow of air is capable of removing the heat from the system.

Extensive research work focusing on designing various hybrid PV/T had been carried out from time to time before it can be mass-produced for commercialization. [6] had carried out a comparison study of four models of PV/T collector. Both single-pass and double-pass system had been evaluated. Heat balance equation had been identified and solved for each model. Performance of a double-pass PV/T solar collector suitable for solar drying system had been investigated [7]. A study on nine different design concepts of combined PV/T water and air solar collector system had been carried out [8]. The design concepts had been divided into four different groups. Two types of PV panel had been investigated, namely, the conventional opaque PV panel and the transparent PV panel. Another study was on two low-cost improvement designs of heat remover placed in a channel of a PV/T system [9]. The improved design involved was done by introducing a thin metal sheet and fin attached to the back wall of the channel.

Thermal efficiency of the double-pass PV/T system with porous media at the lower channel had been evaluated by [10]. Experimental results proved that introducing the porous media at the lower channel has increased the heat transfer area which leads to the increase of the thermal efficiency of the system from 60 to 70 %. A direct coupled outdoor PV/T system in Kerman, Iran, was tested [11]. The design of the system involves the use of a thin aluminum sheet placed in the middle of the air channel as the heat exchanger to cool the PV panels. Another PV/T solar collector with an active cooling system was fabricated to increase the electrical efficiency of a PV module [12].

The objective of this chapter is to evaluate the performance of double-pass and single-pass design of hybrid PV/T. Air base photovoltaic/thermal (PV/T) solar collector combined with various designs of heat exchangers had been studied. For double-pass system, fins and CPC had been used to enhance the performance of the system. As for the single-pass system, to remove the heat from the back of the PV module, five different designs of heat exchanger, honeycomb, stainless steel wool, V-groove, ∇-groove, and rectangular-groove, were installed horizontally into the channel located at the backside of the PV module. Under similar setup of operational condition, each heat exchanger was tested one by one to observe the system performance.

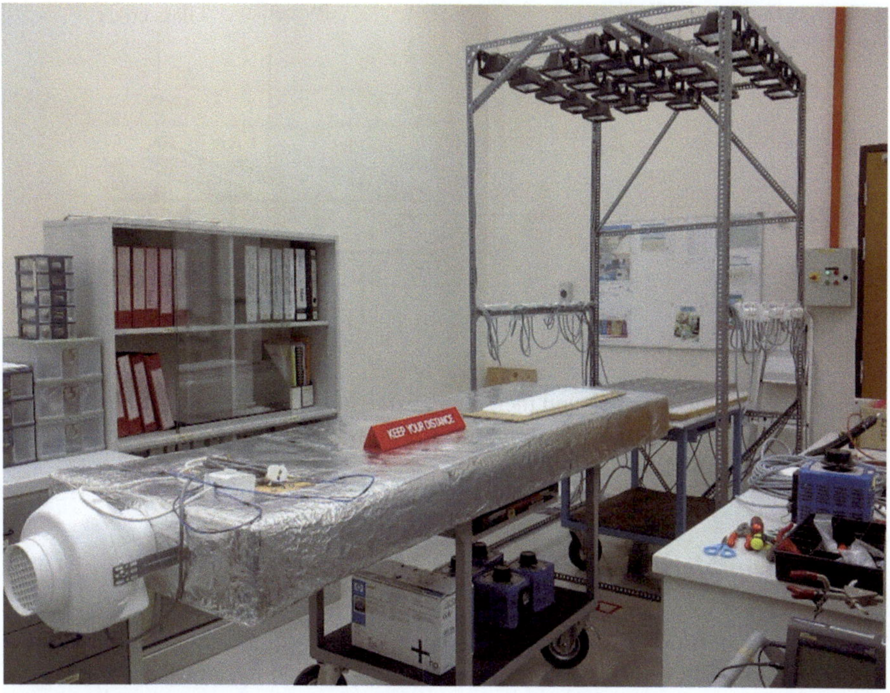

Fig. 5.1 Laboratory fabricated solar simulator

5.2 Designs of PV/T Collector

The performance of each PV/T solar collector had been tested indoor using a laboratory-fabricated solar simulator. Under similar setup of operational condition, the system was tested one by one at irradiance of ~700 to ~800 W/m^2 with mass flow rate span from 0.01 to 0.18 kg/s. Photo in Fig. 5.1 shows the solar simulator used during measurement.

5.2.1 Double-Pass PV/T System with Fins

For a start, a double-pass air base PV/T system with fins as heat exchanger had been fabricated and tested [13]. The collector is shown schematically in Fig. 5.2. The dimension of the collector is 0.85 m × 1.22 m (W × L). The height of the upper channel is 16.5 m. The height of the lower channel can be varied from 30 to 120 mm. Rectangular fins with each size of 2.5 cm high and 120 cm long and a thickness of 0.1 cm were attached along the length of the collector perpendicular to the absorber plate. Air enters through the upper channel formed by the glass cover

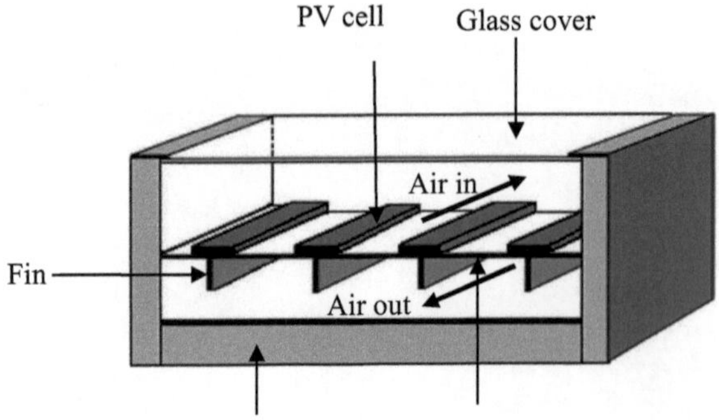

Fig. 5.2 Double-pass PV/T with fins

and the photovoltaic panel, and then it enters the lower channel formed by the back plate and the photovoltaic panel. Figure 5.3 shows photo of the design.

5.2.2 Double-Pass PV/T System with CPC and Fins

In order to improve the performance of the first system, compound parabolic concentrator (CPC) had been introduced to the second system [14]. Figure 5.4 shows the improved system and Fig. 5.5 shows the photo of the system. The size of the system remains the same. The combination with concentrating reflectors has a potential to increase significantly the power production for the same size of solar cell area. The total area covered by solar cells is 0.38 m^2. CPC with concentration of 1.86 is used as the reflector and located parallel to the airflow.

Based on the previous designs of PV/T collector [13] and [14], the electrical efficiency of the collector reduced by nearly 50 % since the PV module is covered by glass plate. Electrical energy is considered as a high-quality source of energy, and reducing its value defeats the purpose of having good quality of PV module. Furthermore, the PV/T design with double-pass system is considered bulky to be further integrated within buildings. Therefore, the next designs were only single-pass PV/T system without glass cover.

5.2.3 Single-Pass PV/T System with ∇-Grooved Absorber Plate

The third design was a single-pass PV/T system with aluminum ∇-grooved absorber plate [15]. The thickness of the aluminum was 0.7 mm, attached to the back of the PV module. Figure 5.6 shows the cross-sectional diagram of the PV/T system.

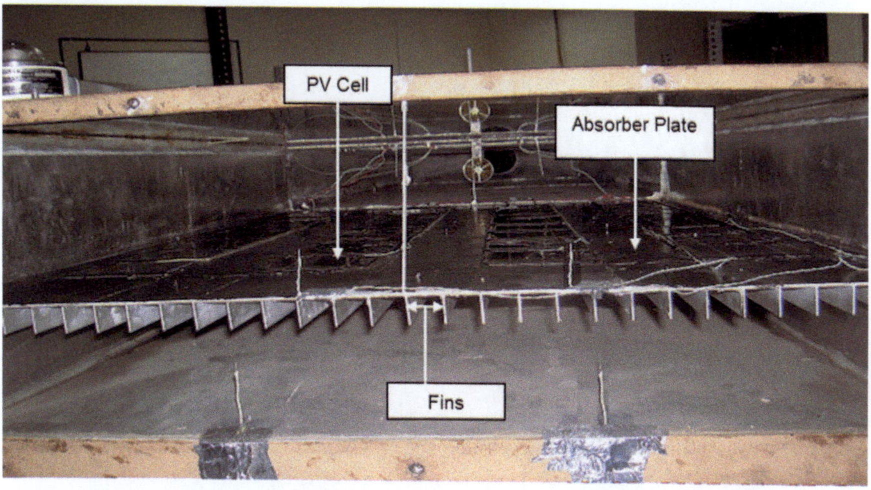

Fig. 5.3 Photo double-pass PV/T system with fins

Fig. 5.4 Double-pass PV/T with fins and CPC

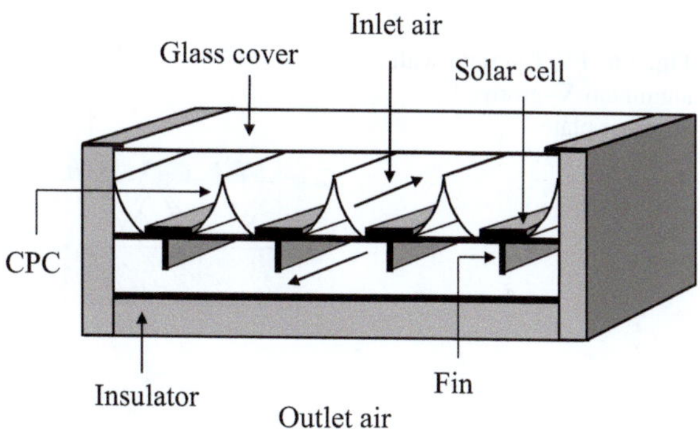

5.2.4 *Single-Pass PV/T System with V-Grooved Absorber Plate*

Figure 5.7 illustrates the fourth design of the PV/T collector with V-groove heat exchanger. V-groove is fabricated using locally purchased aluminum sheet. The thickness of the aluminum sheet is ~0.5 mm. The V-groove was also installed at the back of the PV module using same method as ∇-grooved absorber plate.

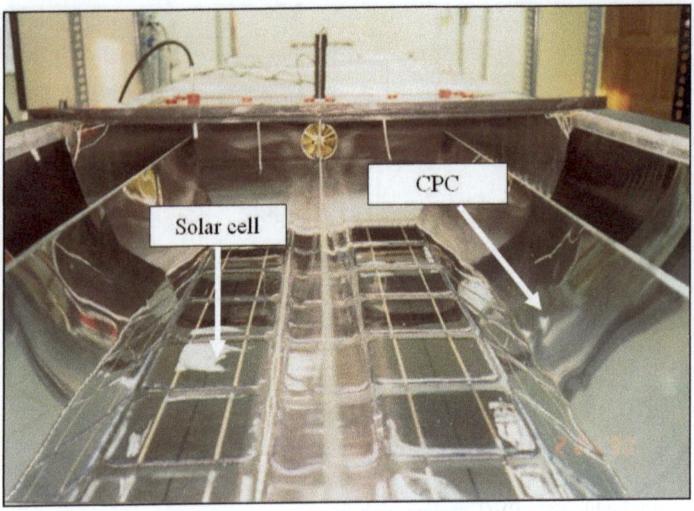

Fig. 5.5 Photo double-pass PV/T system with fins and CPC

Fig. 5.6 PV/T system with aluminum ∇-grooved absorber plate

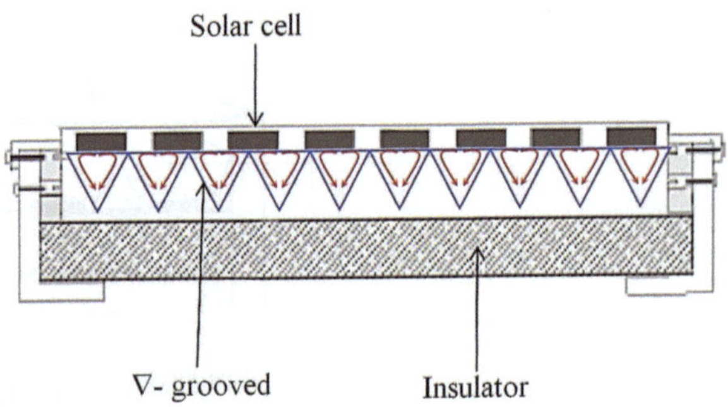

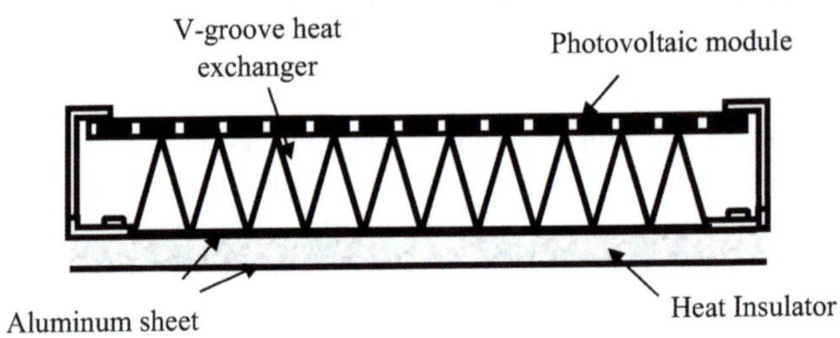

Fig. 5.7 PV/T collector with V-groove heat exchanger

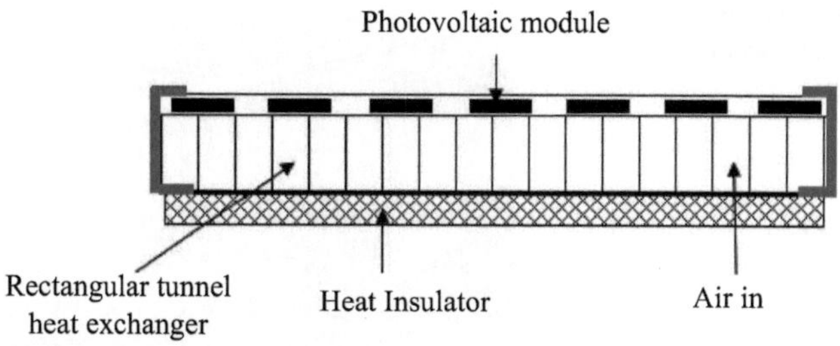

Fig. 5.8 PV/T system with rectangular tunnel heat exchanger

5.2.5 *Single-Pass PV/T System with Rectangular Tunnel Absorber Plate*

The fifth design of heat exchanger is the shape of rectangular tunnel [16]. Rectangular tunnel bars with the size of 1.2 × 2.5 × 120 cm (width × height × length) was installed to the back of the PV module. The material of the rectangular tunnel was an aluminum. Figure 5.8 shows the design of the PV/T system.

5.2.6 *Single-Pass PV/T System with Stainless Steel Absorber Plate*

The next design of the PV/T collector is shown in Fig. 5.9. The heat exchanger chosen was locally purchased stainless steel wool. The stainless steel wool had been tested to prove that it will not rust over time. The stainless steel wool was attached to a piece of aluminum sheet. The complete design was installed to the back of the PV module using the same method as shown in Fig. 5.9.

5.2.7 *Single-Pass PV/T System with Honeycomb Heat Exchanger*

The final design of the PV/T collector is shown in Fig. 5.10 [17] and photo in Fig. 5.11. It is a collector with honeycomb heat exchanger. Locally purchased aluminum sheet had been made into corrugated sheet. The thickness of the sheet is ~0.2 mm. Five pieces of aluminum corrugated sheets were joined together to fabricate a piece of compact honeycomb with hexagonal geometry. The honeycomb was installed horizontally into the channel located at the backside of the PV module.

Fig. 5.9 PV/T collector with stainless steel wool heat exchanger

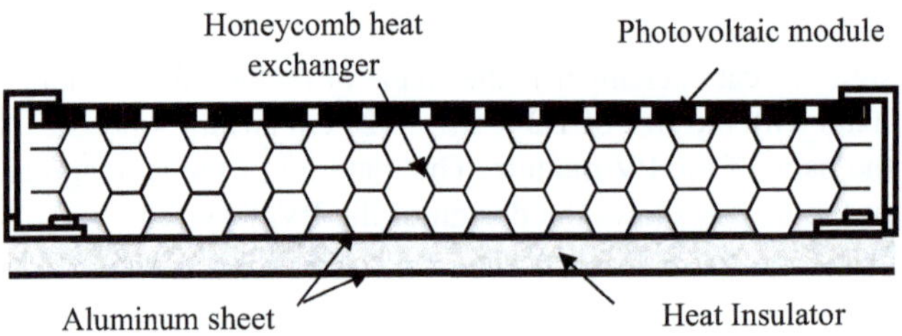

Fig. 5.10 PV/T collector with honeycomb heat exchanger

Fig. 5.11 Photo of PV/T collector with honeycomb heat exchanger

The characteristic of PV/T collector with each design of heat exchanger was studied to identify their overall performance. The performance of the collectors was evaluated and compared in terms of their electrical and thermal efficiencies.

5.3 Experimental Procedures

Heat exchanger with design of honeycomb, V-groove, and stainless steel wool was tested at the National Metrology Laboratory (NML), SIRIM Berhad, Sepang. The double-pass systems, ∇-grooved and rectangular tunnel, were tested at UKM, Bangi. The experiment was run under two similar conditions of irradiance and mass flow rate. At NML-SIRIM the PV/T system was tested indoor at 828 W/m^2 with mass flow rates of air through the system set to 0.021, 0.042, 0.085, 0.110, and 0.128 kg/s. A monocrystalline (SolarWorld, SW 85) photovoltaic module was used to harvest the electrical energy. The thermal system consists of a blower attached into a galvanized ducting with length of 2.5 m, a voltage regulator to control the airspeed of the blower between 0.4 to 1.5 m/s flow through the system, and a heater controlled by a voltage regulator which was installed into the ducting in order to ensure that the inlet temperature is equivalent to the ambient temperature.

At UKM, Bangi, the PV/T system was tested indoor at 817 W/m^2 with mass flow rates of the air through the system set to 0.0110, 0.0287, 0.0409, 0.0552, and 0.0754 kg/s. A solar panel with model SHARP NE-80E2EA was used to harvest the electrical energy. The thermal system consists of a blower attached to a ducting, a voltage regulator to control the airspeed of the blower, and halogen lamps controlled by a voltage regulator to ensure that the inlet temperature is equivalent to the ambient temperature.

The double-pass systems with fins and with CPC and fins were also tested at UKM, Bangi. The PV/T system was tested indoor at 700 W/m^2 with mass flow rates of the air through the system set to 0.027 to 0.16 kg/s. A monocrystalline solar cell, Solar Tech SC2240, was used to harvest the electrical energy.

Type-T thermocouples were placed at various locations of the system. Inlet temperature, panel temperature, outlet temperature, and temperature at the back of the thermal insulator were measured and used to calculate the thermal efficiency of the system. At each mass flow rate setting, time of not less than 90 min was allowed for temperature stabilization. For the first step, the PV/T collector system efficiency was studied without any heat exchanger installed into the system. Then the system efficiency was tested with heat exchanger one by one separately.

5.4 Results and Discussion

All parameters involved in the calculation of thermal and electrical efficiency were measured in details. Maximum current, I_m (A), and maximum voltage, V_m (V), were determined from I/V curve measurement of the PV module using digital multimeter and rheostat. A_c is the area of the solar cell and S is the average irradiance value during the experiment. Equation (5.1) is used to calculate the electrical efficiency, π_{el}, of a PV/T system. Measurements of short-circuit current, I_{sc} (A), and open-circuit voltage, V_{oc} (V), can be done by direct connection between the multimeter and the PV module.

$$\pi_{el} = \frac{I_m \text{ x } V_m}{\text{A}_\text{c} \text{ x S}} \text{ x } 100\% \tag{5.1}$$

Inlet temperature, T_i; panel temperature, T_s; outlet temperature, T_o; and temperature at the back of the thermal insulator, T_b, were measured and used to calculate the thermal efficiency of the system. Air velocity, V_{av}, at input location was calculated. Mass flow rate, $\dot{m}$, of the airflow through the system was calculated using Eq. (5.2).

$$\dot{m} = AV_{av} \tag{5.2}$$

Equation (5.3) was used to calculate the thermal efficiency, π_{th}, of the developed PV/T system. C_p is the specific heat of air, and A_p is the area of the PV/T collector:

$$\pi_{th} = \frac{mC_p(T_o - T_i)}{A_p S} \text{ x } 100\% \tag{5.3}$$

Using Eqs. (5.1) and (5.3), electrical and thermal efficiency for all design were calculated and plotted in graph form for further analysis. Graph in Fig. 5.12 shows the plot of the thermal efficiency for double-pass PV/T collector with fins. Maximum thermal efficiency is 68 % at 0.016 kg/s and irradiance of 700 W/m^2. Figure 5.13 for electrical efficiency for the system. Overall electrical efficiency is 2 % for the whole range of mass flow rate.

Graph in Fig. 5.14 shows the plot of the thermal efficiency for double-pass PV/T collector with CPC and fins. Maximum thermal efficiency is 63 % at 0.011 kg/s and irradiance of 700 W/m^2. Figure 5.15 is the plot of electrical efficiency for the system. Maximum electrical efficiency is 1.55 %. Figure 5.16 illustrates how the power of the PV module can be improved with the usage of CPC. The maximum current is 0.35 A with fins only and 0.6 with CPC and fins, improved by 0.25 A. The maximum power for PV/T with fins only is 3.0 W and with CPC and fins 5.7 W, improved by 2.7 W. The improvement is very significant.

Graph in Fig. 5.17 shows the plot of the thermal efficiency for collector with ∇-grooved heat exchanger. The thermal efficiency, π_{th}, increases with the increase of mass flow rate. The maximum thermal efficiency, π_{th}, is ~75.0 % and without heat exchanger is 40.0 %. Graph in Fig. 5.18 shows the plot of for electrical efficiency, π_{el}, for the system. Overall, collectors show an increasing trend for electrical efficiency, π_{el}, with the increase of the fluid mass flow rate. The electrical efficiency improve by 2.0 % with the usage of ∇-grooved heat exchanger.

Graph in Fig. 5.19 shows the plot of the thermal efficiency for collector with V-groove heat exchanger. Result shows that the thermal efficiency, π_{th}, increases with the increase of mass flow rate. The maximum thermal efficiency, π_{th}, is ~75.0 % and without is only 27.0 %. Graph in Fig. 5.20 shows the plot of for electrical efficiency, π_{el}, for the system. The difference of electrical efficiency for collectors with and without heat exchanger is not so significant.

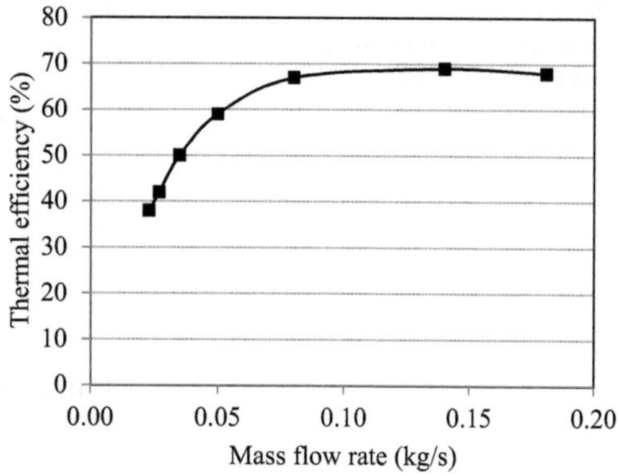

Fig. 5.12 Thermal efficiency curve for double-pass PV/T with fins

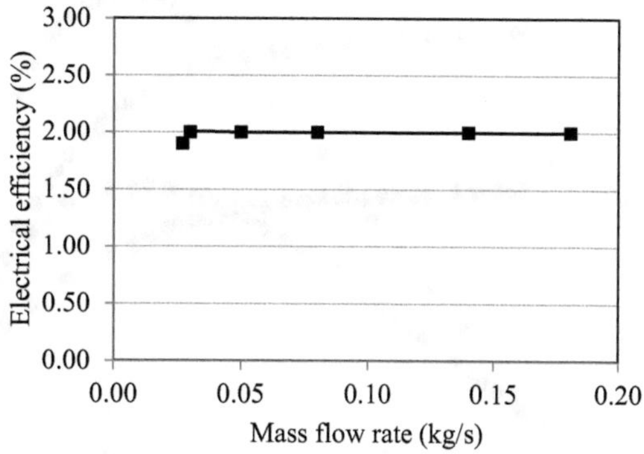

Fig. 5.13 Electrical efficiency curve for double-pass PV/T with fins

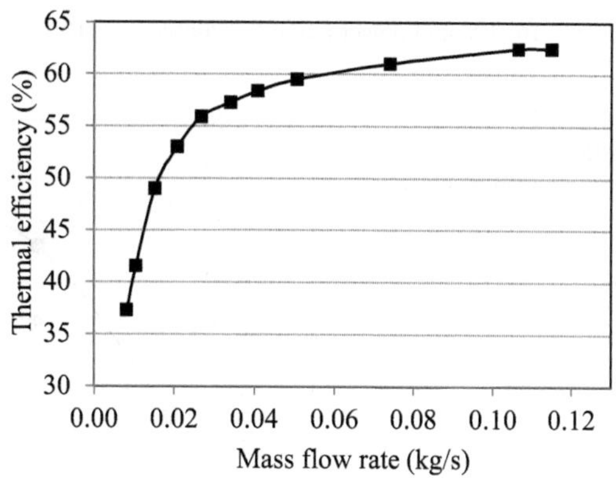

Fig. 5.14 Thermal efficiency curve for double-pass PV/T with CPC and fins

Graph in Fig. 5.21 shows the plot of the thermal efficiency for collector with rectangular tunnel heat exchanger. The thermal efficiency, π_{th}, increases with the increase of mass flow rate. The maximum thermal efficiency, π_{th}, is ~75.0 % and

Fig. 5.15 Electrical efficiency curve for double-pass PV/T with CPC and fins

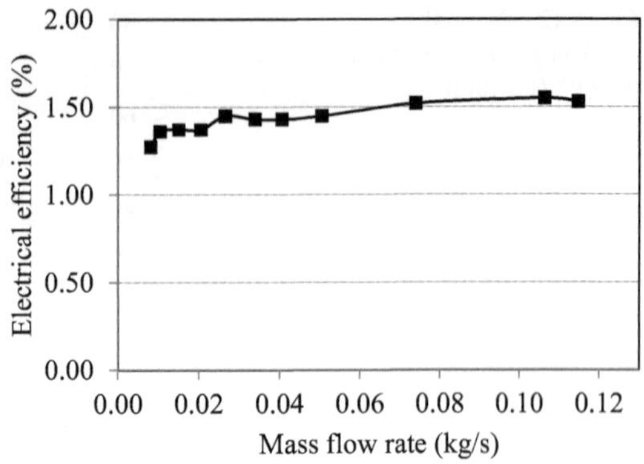

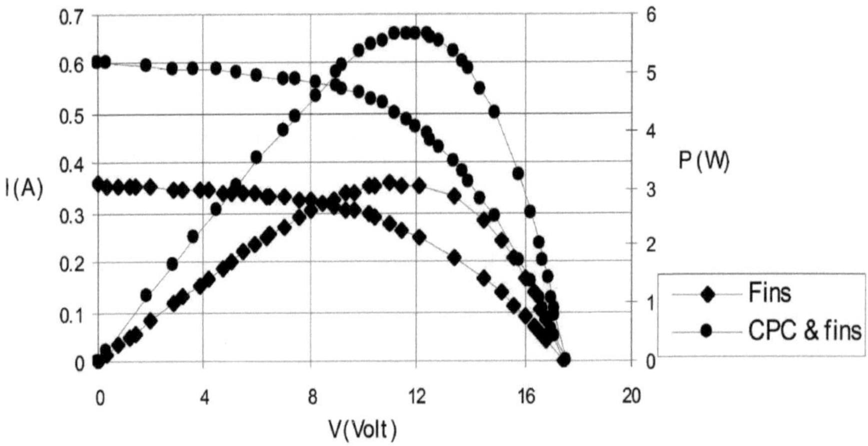

Fig. 5.16 I/V and power curve for double-pass PV/T with fins only and with CPC and fins

Fig. 5.17 Thermal efficiency curve for ∇-grooved heat exchanger

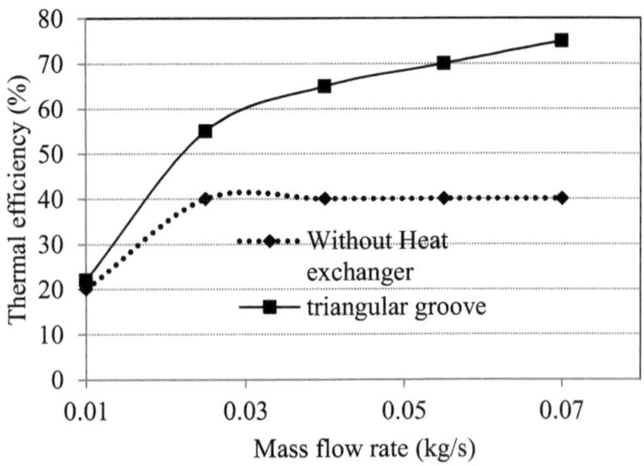

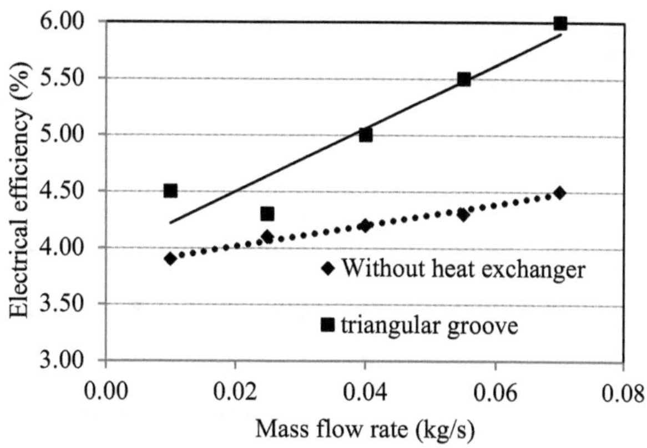

Fig. 5.18 Electrical efficiency curve for ∇-grooved heat exchanger

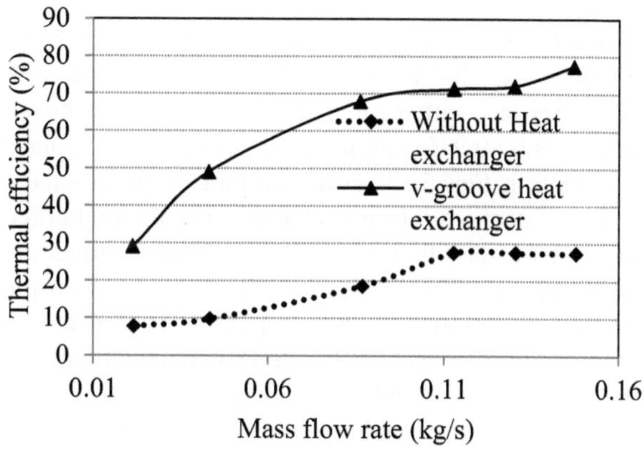

Fig. 5.19 Thermal efficiency curve for V-groove heat exchanger

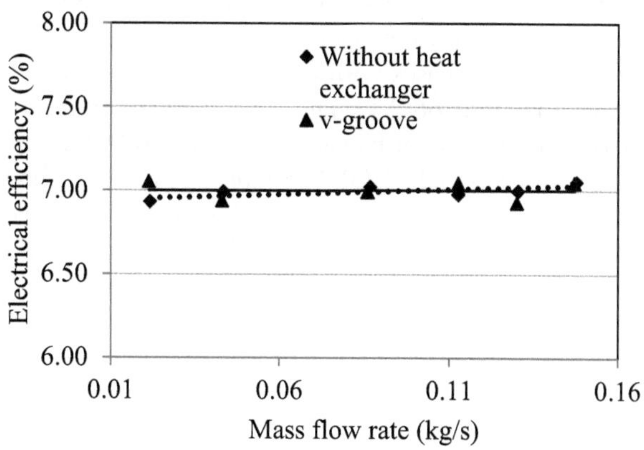

Fig. 5.20 Electrical efficiency curve for V-groove heat exchanger

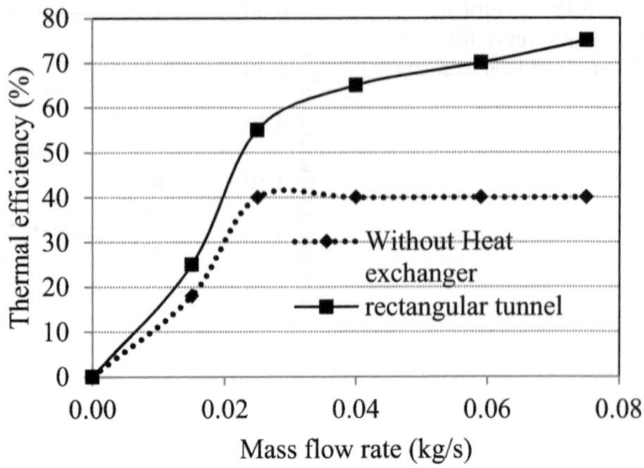

Fig. 5.21 Thermal efficiency curve for rectangular tunnel heat exchanger

without heat exchanger is 40.0 %. Graph in Fig. 5.22 shows the plot of for electrical efficiency, π_{el}, for the system. The collectors show an increasing trend for electrical efficiency, π_{el}, with the increase of the fluid mass flow rate. The electrical efficiency improves by 0.2 % with the usage of triangular tunnel heat exchanger.

Graph in Fig. 5.23 shows the plot of the thermal efficiency for collector with stainless steel wool heat exchanger. The thermal efficiency, π_{th}, increases with the increase of mass flow rate. The maximum thermal efficiency, π_{th}, is ~87.0 % and without is only 27.0 %. Graph in Fig. 5.24 shows the plot of for electrical efficiency, π_{el}, for the system. Usage of stainless steel heat exchanger drops the electrical efficiency collectors by 0.1 % compared to without heat exchanger. The structure of the stainless steel wool is not uniformed to allow air to flow freely through it from inlet to outlet. It caused a pressure drop to the flowing air and is not suitable to be used to remove heat from the back of the PV module.

Graph in Fig. 5.25 shows the plot of the thermal efficiency for collector with honeycomb heat exchanger. The experimental result shows that the thermal efficiency, π_{th}, increases with the increase of mass flow rate up to 0.11 kg/s. Above mass flow rate of 0.11 kg/s, the thermal efficiency, π_{th}, remains stable. The maximum thermal efficiency, π_{th}, of the collector with the honeycomb is ~87.0 % and without is only 27.0 %. Graph in Fig. 5.26 shows the plot of the electrical efficiency, π_{el}, for collector with honeycomb heat exchanger. Overall, collectors show an increasing trend for electrical efficiency, π_{el}, with the increase of the fluid mass flow rate. The electrical efficiency improves by 0.2 % with the usage of honeycomb heat exchanger.

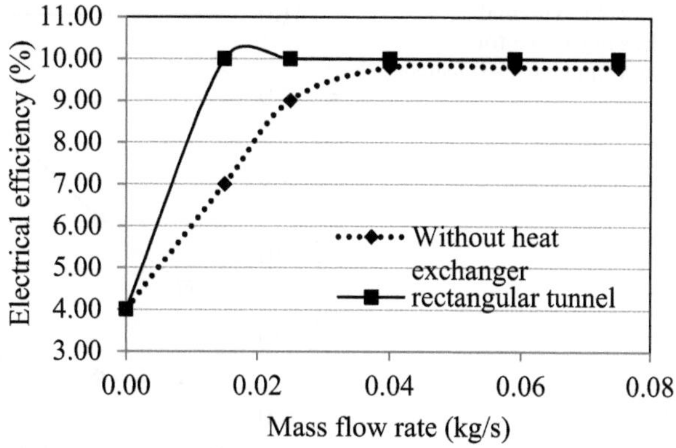

Fig. 5.22 Electrical efficiency curve for rectangular tunnel heat exchanger

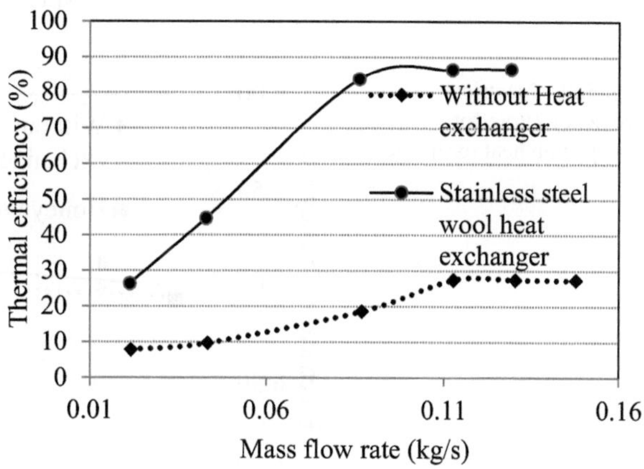

Fig. 5.23 Thermal efficiency curve for stainless steel wool heat exchanger

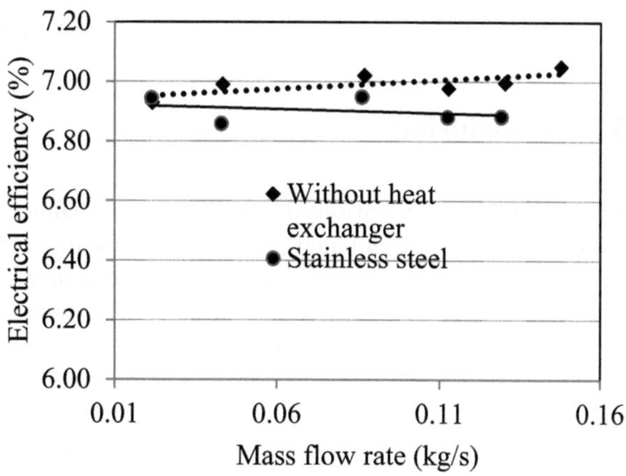

Fig. 5.24 Electrical efficiency curve for stainless steel wool heat exchanger

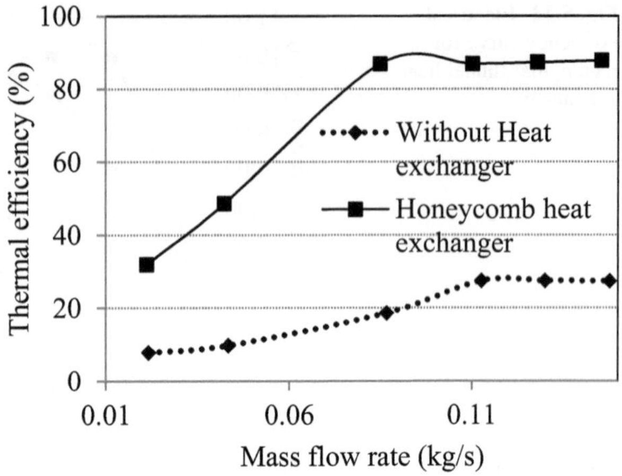

Fig. 5.25 Thermal efficiency curve for honeycomb heat exchanger

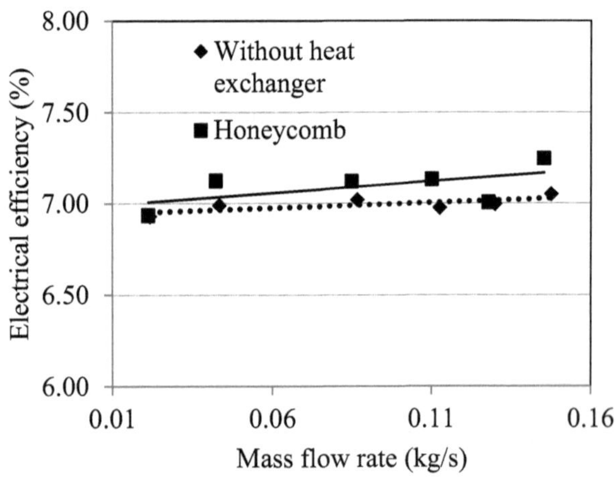

Fig. 5.26 Electrical efficiency curve for honeycomb heat exchanger

5.5 Conclusion

The performance for various designs of heat exchanger for hybrid single-pass PV/T collector had been evaluated and explained in this chapter. Every design had its own capability to harvest solar energy and convert it to thermal energy and electrical energy. Overall, PV/T designs with heat exchangers show improved efficiency for both electrical and thermal energy, except for PV/T with stainless steel wool which showed a decrease of electrical efficiency due to its nonuniform structure. The use of PV/T in combination with CPC has the potential to significantly increase the power production from a given solar cell area.

Fins attached to the back of the photovoltaic panel in the double-pass system increase the heat transfer to the moving air. Since the airflow is interrupted, turbulence is introduced. This turbulence increases the heat transfer coefficient.

The fins also provide augmented heat transfer area. The combined effect of these two phenomena will improve the heat transfer from the plate to the flowing air.

PV/T with honeycomb heat exchanger is capable to increase thermal energy by ~50 % and electrical energy by 0.2 %. PV/T with V-groove is capable to enhance thermal energy by ~40 % and no significant improvement for electrical efficiency. PV/T with stainless steel shows ~50 % thermal energy improvement. PV/T with ∇-grooved heat exchanger shows increasing value of ~35 % for thermal energy and 2.0 % for electrical energy. Finally PV/T with triangular tunnel heat exchanger shows increasing value of ~35 % for thermal energy and 0.2 % for electrical energy. In the future, findings from this experiment can be further investigated to be applied in solar drying system and space heating.

References

1. M.A. Green, *Solar Cells, Chapter 11* (Prentice-Hall, Englewood Cliffs, NJ, 1982)
2. M.A. Hasan, K. Sumathy, Photovoltaic thermal module concepts and their performance analysis: A review. Renew. Sust. Energ. Rev. **14**, 1845–1859 (2010)
3. R. Kumar, M.A. Rosen, A critical review of photovoltaic–thermal solar collectors for air heating. Appl. Energy **88**, 3603–3614 (2011)
4. G.N. Tiwari, R.K. Mishra, S.C. Solanki, Photovoltaic modules and their applications: A review on thermal modeling. Appl. Energy **88**, 2287–2304 (2011)
5. F. Hussain, M.Y.H. Othman, K. Sopian, B. Yatim, H. Ruslan, H. Othman, Design development and performance evaluation of photovoltaic/thermal (PV/T) air base solar collector. Renew. Sustain. Energy Rev. **25**(431–441) (2013)
6. A.A. Hegazy, Comparative study of the performances of four photovoltaic/thermal solar air collectors. Energ. Convers. Manage. **41**, 861–881 (2000)
7. K. Sopian, H.T. Liu, S. Kakac, T.N. Veziroglu, Performance of a double pass photovoltaic thermal solar collector suitable for solar drying systems. Energ. Convers. Manage. **41**, 353–365 (2000)
8. H.A. Zondag, D.W.D. Vries, W.G.J. van Helden, R.J.C. van Zolingen, A.A. van Steenhoven, The yield of different combined PV-thermal collector designs. Sol. Energy **74**, 253–269 (2003)
9. Y. Tripanagnostopoulos, Aspects and improvements of hybrid photovoltaic/thermal solar energy systems. Sol. Energy **81**, 1117–1131 (2007)
10. K. Sopian, M.A. Alghoul, E.M. Alfegi, M.Y. Sulaiman, E.A. Musa, Evaluation of thermal efficiency of double-pass solar collector with porous–nonporous media. Renew. Energy **34**, 640–645 (2009)
11. A. Shahsavar, M. Ameri, Experimental investigation and modeling of a direct-coupled PV/T air collector. Sol. Energy **84**, 1938–1958 (2010)
12. H.G. Teo, P.S. Lee, M.N.A. Hawlader, An active cooling system for photovoltaic modules. Appl. Energy **90**, 309–315 (2012)
13. M.Y. Othman, B. Yatim, K. Sopian, M.N.A. Bakar, Performance studies on a finned double-pass photovoltaic-thermal (PV/T) solar collector. Desalination **209**, 43–49 (2007)
14. M.Y. Othman, B. Yatim, K. Sopian, M.N.A. Bakar, Performance analysis of a double-pass photovoltaic/thermal (PV/T) solar collector with CPC and fins. Renew. Energy **30**, 2005–2017 (2005)
15. M.Y.H. Othman, H. Ruslan, K. Sopian, G.L. Jin, Performance study of photovoltaic-thermal (PV/T) solar collector with ∇-grooved absorber plate. Sains Malays. **38**(4), 537–541 (2009)

16. G.L. Jin, A. Ibrahim, Y.K. Chean, R. Daghigh, H. Ruslan, S. Mat, M.Y. Othman, K. Ibrahim, A. Zaharim, K. Sopian, Evaluation of Single-Pass Photovoltaic-Thermal Air Collector with Rectangle Tunnel Absorber. Am. J. Appl. Sci. **7**, 277–282 (2010)
17. F. Hussain, M.Y. Othman, B. Yatim, H. Ruslan, K. Sopian, Z.Anuar, S. Khairuddin, Performance of a single pass air base photovoltaic/thermal solar collector with and without hexagonal honeycomb heat exchanger, World Renewable Energy Forum, WREF 2012, Including World Renewable Energy Congress XII and Colorado Renewable Energy Society (CRES) Annual Conference, Vol. 2, (2012), pp. 859–864

Chapter 6
Teaching Photovoltaic Principles at the University

Hussein A. Kazem

6.1 Introduction

This chapter presents teaching photovoltaic principles at universities. To understand the principles, design, operation, and performance of photovoltaic systems, it is important to discuss the solar radiation and different elements that affect them. The first part of this chapter discusses solar radiation, insolation, elements that affect solar radiation, insolation on tracking surface, stationary surface, and horizontal surface.

The second part of this chapter talks about the photovoltaic (henceforth PV) history, structure, types, performance parameters, operation, and design. Elements that affect PV are going to be discussed. At the end of this chapter, the full picture of the PV operation and performance is going to be clarified. The following points will be dealt with: solar radiation from the sun, radiation through atmosphere, radiation observed by PV, electricity generated in PV, and electricity received by load.

6.2 Solar Radiation and Insolation

Solar radiation can be defined as the beam emitted by the sun. The maximum solar radiation (flux density) that reaches the earth's surface is about 1 kW/m2 in a wavelength band from 0.3 to 2.5 μm. This is called shortwave radiation which includes the visible spectrum. The outward radiation energy fluxes emitted by the earth are also of the order of 1 kW/m2, but occur in an infrared wavelength band from 5 to 25 μm, and this is called long-wave radiation. The solar radiation is

H.A. Kazem, PhD (✉)
Sohar University, P O Box 44, Sohar 311, Oman
e-mail: h.kazem@soharuni.edu.om

A. Sayigh (ed.), *Photovoltaics for Sustainable Electricity and Buildings*,
DOI 10.1007/978-3-319-39280-6_6

classified into two types: extraterrestrial solar radiation and global solar radiation [1, 2]. The extraterrestrial solar radiation is the solar radiation outside the atmosphere or in the space. The extraterrestrial solar radiation, G_{extra}, is given by

$$G_{extra} = I_o \left[1 + 0.034 \cos\left(\frac{2\pi d}{365}\right) \right] \tag{6.1}$$

where I_o is the solar constant, 1367 W/m2, and d is the number of days.

The power density of solar radiation in space is higher than the one on the ground due to atmospheric absorption. The upper atmosphere absorbs radiation of frequencies above 1000 THz (300 nm) producing photo-ionization, causing photochemical reactions and generally heating up the air. Near 25 km altitude, the ozone layer absorbs much of it. The ozone is amazingly opaque to ultraviolet. Although several different mechanisms are the reason behind generating solar radiation, the bulk of it is of the black-body type. The energy per unit volume per unit frequency interval inside a hollow isothermal black body is given by Plank's law:

$$\frac{dW}{df} = \frac{8\pi h}{c^3} \frac{f^3}{\exp(hf/kT) - 1} \quad J.m^{-3}.Hz^{-1} \tag{6.2}$$

Energy flux is, as we stated, the same as power density, P:

$$\begin{aligned} \frac{dP}{df} &\propto \frac{f^3}{\exp(hf/kT) - 1} \quad WJ.m^{-2}.Hz^{-1} \\ \frac{dP}{df} &\propto \frac{f^3}{\exp(hf/kT) - 1} \quad WJ.m^{-2}.Hz^{-1} \end{aligned} \tag{6.3}$$

In terms of wavelength, $\frac{dP}{d\lambda} \propto \frac{\lambda^{-5}}{\exp(hf/kT\lambda)-1}$ $WJ.m^{-2}$ *per m of wavelength interval,* global solar radiation is the total solar radiation that reaches the earth's surface. The global solar radiation on a tilt surface consists of three components [3, 4], namely, direct solar radiation, G_B; diffuse solar radiation, G_D; and reflected solar radiation, G_R, as shown in Fig. 6.1.

From Fig. 6.1, the global radiation which falls on a tilt surface is given by

$$G_T = G_B + G_D + G_R \tag{6.4}$$

However, on a horizontal surface, $G_R = 0$, and therefore, the amount of global solar radiation is given by

$$G_T = G_B + G_D \tag{6.5}$$

The amount of global solar radiation depends on the sky clearness index, K_T, which is given by

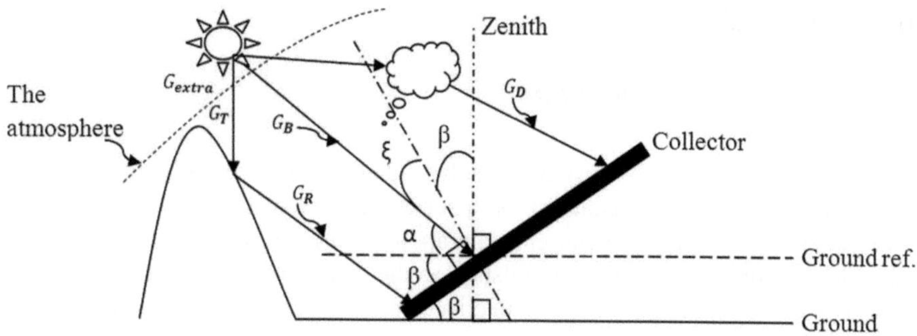

Fig. 6.1 Solar radiation components on a tilt surface

$$\frac{G_T}{G_{extra}} = K_T \tag{6.6}$$

where K_T has a value in the range of 0–1. Here, if the sky is overcast, then little amount of solar radiation is received and vice versa [2].

The diffuse solar radiation is the amount of the solar radiation which is being scattered by clouds and sky impurities. The diffuse solar radiation index K_D is given by

$$\frac{G_D}{G_T} = K_D \tag{6.7}$$

The reflected solar radiation is estimated by using

$$G_R = G_T \rho \tag{6.8}$$

where $\boldsymbol{\rho}$ is the ground albedo, and it is usually of values 0.1 for soil, 0.2 for sand, and 0.3 for grass.

Solar energy is the amount of solar radiation multiplied by time. Therefore, the daily global solar energy on a horizontal surface is the "average of the daily global solar radiation multiplied by the length of the solar day" which is the time from sunrise to sunset. The solar day length (S_o) is calculated by using

$$S_o = \frac{2}{15} \cos^{-1}(-\tan L . \tan DEC) \tag{6.9}$$

where L is the latitude and DEC is the declination angle. Based on the solar day length, the extraterrestrial solar energy (E_{extra}), global solar energy (E_T), beam solar energy (E_B), and direct solar energy (E_D) can be determined by multiplying G_{extra}, G_T, G_B, and G_D with S_o, respectively.

6.3 Insolation

6.3.1 *Geometry of the Sun*

Insolation is the "power density of the solar radiation." In Sect. 6.2, solar constant has been defined as the "insolation on a surface that faces the sun outside earth's atmosphere." The value of solar constant approximately is 1360 $W\backslash m^2$. Also, it is suitable to define a surface solar constant as the "value of insolation on a surface that, at sea level, faces the vertical sun on a clear day." The convenient value of the "constant" is about 1000 $W\backslash m^2$ or "one sun." At non-vertical situation of the sun, due to the larger air mass through which the solar rays have to pass, the insolation is likewise smaller.

The insolation depends on [1]:

1. The transparency of the atmosphere
2. Surface-relative orientation with respect to the sun

The main sources of errors are:

1. It was assumed that the period between successive sunrises throughout the year is constant. That is not the accurate situation, so the "equation of time" will be used for correction, which is going to be discussed in the next Sect. 6.3.2.
2. The "mean local time" has been used in different formulas, which differs from the civil time (the time measured at the center of each time zone).

Effects 1 and 2 could be easily corrected by introducing the "time offset" (see Sect. 6.3.2).

3. The geometry of the situation (only) has been considered and used in the formulas. Because of the diffraction of the light, which happened due to the presence of the atmosphere, the sun is still visible when it is fairly below the geometric horizon. This causes the apparent sunrise to be earlier than the geometric one and the apparent sunset to be later. Using a solar zenith angle at sunrise and sunset of 90.833 instead of the geometric 90 could correct this effect.
4. Perfectly transparent atmosphere has been assumed during the insolation data collection. Meteorological conditions do, of course, alter in a major way the amount of useful sunlight.

The south to north location of a point on the earth is identified by latitude *L*. The west to east location of a point on the earth is identified by longitude *LOD*, by measuring the angular distance from the Greenwich meridian (or prime meridian, where longitude is 0) along the equator.

However, the sun position in the sky can be defined by many angles such as hour angle *HA*, angle of declination *DEC*, solar altitude *SOLAL*, solar azimuth *SOLAZM*, zenith angle *SOLZN*, and incidence angle *SOLIN*. The zenith angle is defined as "the angle between the local vertical and the line from the observer to the sun," and the azimuth angle is "the azimuth from north of the sun's beam projection on the

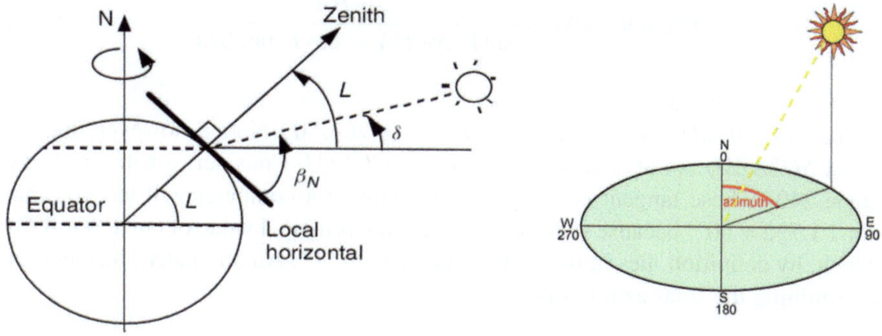

Fig. 6.2 Solar zenith angle: the angle between the local zenith (*up*) and the line of sight to the sun

horizontal plane clockwise positive." This is a topocentric system—the observer is at the origin of the coordinates. Two different points of view have been used: a geocentric system (origin at the center of earth) and a heliocentric system (origin at the center of the sun). Both *SOLZN* and *SOLAZM* are functions of the day of the year, *d*, and the local time of day, *t*, and the latitude of the observer *L* in our topocentric system (Fig. 6.2).

Observe that the time, ***t***, in the formulas is different from the time in our watch. These times differ by the time offset which has two components, one related to the difference due to the equation of time *EOT* and one owing to the longitude of the place of interest from that of the center of the time zone [5].

Hour angle, *HA*, is represented by the time of day, a usage borrowed from astronomers who mostly worked in the past at night and thus preferred to count a new day from noon rather than from midnight. They defined the hour angle as

$$HA \equiv \frac{360}{24}(t-12) \quad \text{Degree } (t \text{ in hours, 24-h clock}) \tag{6.10}$$

$$\text{HA} \equiv \frac{2\pi}{86400}(t-43200) \quad \text{Radians } (t \text{ in hours, 24-h clock}) \tag{6.11}$$

The day of the year or "season" is represented by the solar declination, *DEC*, by the latitude of the sun:

$$DEC = 23.44 \sin\left[360\left(\frac{d-80}{365.25}\right)\right] \quad \text{degrees} \tag{6.12}$$

The solar zenith angle and the solar azimuth are given by [6]

$$\cos \text{SOLZN} = \sin DEC.\sin L + \cos DEC.\cos L.\cos HA \tag{6.13}$$

$$\tan \text{SOLAZM} = \frac{\sin HA}{\sin L.\cos HA - \cos L.\tan DEC} \tag{6.14}$$

where L is the latitude of the observer.

To find *SOLAZM*, we have to take $tan^{-1}(tan\ SOLAZM)$. Notice, however, that $tan^{-1}(tan\ SOLAZM)$ is not necessarily equal to *SOLAZM*. Consider, for instance, the angle 240° whose tangent is 1.732. A calculator or a computer will tell you that tan-1 1.732 = 60°° because such devices yield the principal value of tan-1 *SOLAZM* which, by definition, lies in the range from -90o to 90°. An alternative formula for determining the solar azimuth is

$$\cos(180 - SOLAZM) = -\frac{\sin L.\cos SOLZN - \sin DEC}{\cos L.\sin SOLZN} \tag{6.15}$$

At both sunset and sunrise, $SOLZN = 90^o$; thus, $cos\ (SOLZN) = 0$. From Eq. 6.13,

$$\cos HA_R = \cos HA_S = -\tan DEC.\tan L \tag{6.16}$$

where $HA_{R,S}$ is the hour angle at either sunrise or sunset. The hour angle HA_S, at sunset, is positive and HA_R, at sunrise, is negative:

$$HA_R = -HA_S \tag{6.17}$$

6.3.2 Time Zones and Time Offset

The "local mean solar time" is not a practical measure of time for everyday use since it depends on the observer longitude. It varies by 1 h for every 15° of longitude. The use of time zones, 1 h or 15° wide, circumvents this difficulty. Regardless the position of the observer in each zone, the time is the same. At the zone boundaries, the time changes abruptly by 1 h. The center meridian of any time zone is a multiple of 15°; the first zone is squarely centered on the zeroth meridian that of Greenwich, and the time there is called "Greenwich mean time," *GMT* (or, to astronomers, "universal time," *UT*).

The time zone is called "standard time" (such as, for instance, PST, for Pacific Standard Time, the -8 time zone centered at 120° W) [7].

The "true solar time," t_{true}, at any given longitude, *LOD*, can be found from

$$t_{true} = t_{local\ mean} + t_{offset} \tag{6.18}$$

where t_{local} and t_{true} are expressed in hours and minutes but t_{offset} is in minutes only

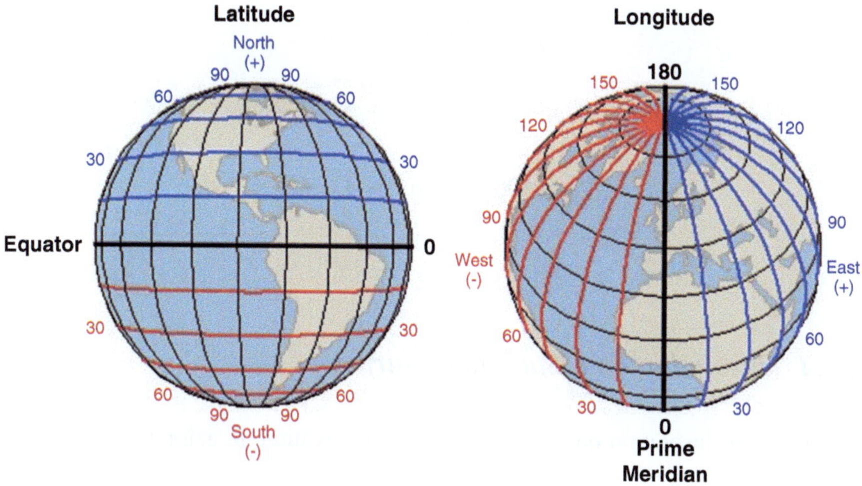

Fig. 6.3 Longitude and latitude

$$t_{offset} = EOT - 4LOD + 60t_{zone}minutes; \tag{6.19}$$

where LOD is the longitude in degrees ($east > 0$, $west < 0$), t_{zone} is the number of hours of the local time zone referred to the UT (east > 0, west < 0), and EOT is the equation of time (in minutes). The difference between the mean time and the apparent solar time is called equation of time—can be expressed mathematically as

$$\begin{aligned} EOT = 2.292(&0.0075 + 0.1868\cos DD - 3.2077\sin DD \\ &- 1.4615\cos 2DD - 4.089\sin 2DD) \end{aligned} \tag{6.20}$$

where the angle $DD = 2\pi(d-1)/365$ is in radians (Fig. 6.3).

6.3.3 Insolation on a Sun-Tracking Surface

Daily average insolation is

$$< p >= \frac{1}{T}\int_{t_R}^{t_S} P_S.dt \quad \mathrm{W.m^{-2}} \tag{6.21}$$

Assuming (unrealistically) that P_S, which is the solar power density, is a constant from sunrise to sunset, the average insolation, $<p>$, in terms of the hour angle, is

$$< p >= \frac{1}{\pi} HA_S.P_S \;\; W.m^{-2} \tag{6.22}$$

At the equinoxes, $DEC = 0$, and, therefore, $HA_S = \pi/2$, and

$$< p >= \frac{1}{2} P_S \approx 500 \quad W.m^{-2} \approx 43.2 MJ.m^{-2}.day^{-1} \tag{6.23}$$

6.3.4 Insolation on a Stationary Surface

Instantaneous insolation on surfaces with tilt or elevation, β, azimuth, ζ, is

$$P = P_S[\cos\beta.\cos SOLZN + \sin\beta.\sin SOLZN.\cos(SOLAZM - \zeta)] \tag{6.24}$$

The elevation angle is always taken as positive as shown in Fig. 6.4. It is worth mentioning here to check if the sun is shining on the back or the front of the surface. The latter condition would result in a negative sign in the second term inside the brackets:

$$P = \frac{1}{T} \int_{t_R}^{t_S} P dt = \frac{1}{2\pi} \int_{HA_R}^{HA_S} P dHA \tag{6.25}$$

6.3.5 Insolation on Horizontal Surfaces

For horizontal surfaces $\beta = 0$, Eq. 6.24 is reduced to

$$P = P_S \cos SOLZN \tag{6.26}$$

Consequently,

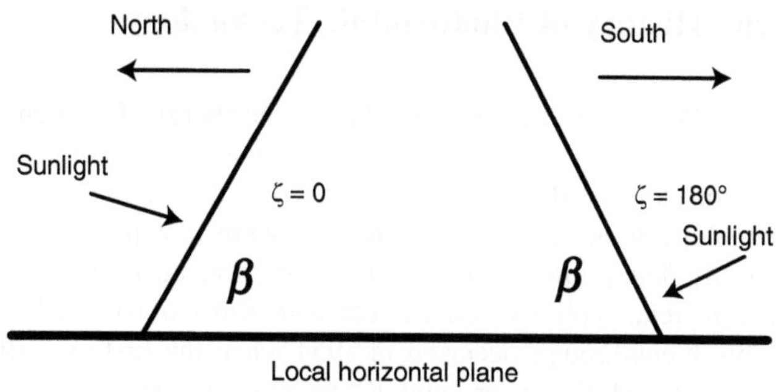

Fig. 6.4 Two surfaces with the same elevation but different azimuths

$$\begin{aligned} P &= \frac{1}{2\pi}\int_{HA_R}^{HA_S} P_S \cos SOLZN \\ &= \frac{P_S}{2\pi}[\sin DEC.\sin L(HA_S - HA_R) + \cos DEC.\cos L(\sin HA_S - \sin HA_R)] \\ &= \frac{P_S}{2\pi}\cos DEC.\cos L(2\sin HA_S + 2HA_S \tan DEC.\tan L) \\ &= \frac{P_S}{\pi}\cos DEC.\cos L(\sin HA_S - HA_S.\cos HA_S) \end{aligned} \tag{6.27}$$

At the equinoxes, $DEC = 0$, $HA_S = \pi/2$; therefore

$$< P >= \frac{P_S}{\pi}\cos L \tag{6.28}$$

At the equator, regardless of DEC, $HA_S = \pi/2$; therefore

$$< P >= \frac{P_s}{\pi}\cos DEC \tag{6.29}$$

6.4 Solar Cell and Photovoltaic

PV is a device that converts light energy directly into electricity and commonly known as "solar cells." The simple example is the solar cell-powered small calculators. Large size and more complicated PV systems provide a large portion of electricity for cities. PV represents one of the most promising means of maintaining our energy intensive standard of living while not contributing to global warming and pollution.

6.5 A Brief History of Photovoltaic Technology

In a solar cell or PV cell, a semiconductor absorbs the energy of sunlight in the form of photons, which are then converted to a voltage via the movement of electrons. The unidirectional flow of these electrons across the cell creates a direct current, DC. PV technology is actually old. Solar cells were first produced in 1883 by Charles Fritts [8]. The photovoltaic effect was discovered in the late 1830s. An explanation of the related photoelectric effect made Einstein win the Nobel Prize in 1921 [9]. A big breakthrough occurred in 1954 when the first crystalline silicon solar cell was developed. Four years later, the technology was used to power a space satellite. The good news was that it worked; the bad news was that the solar cells kept working for few years when NASA did not expect so that the satellite kept sending data to earth when it was no longer needed and took up bandwidth. In the mid-1970s, because of the oil crisis and OPEC, more attention and money were given to find cheaper and new ways of making solar cells. One development of this research push was the first amorphous silicon cell developed by Carlson and Wronski [10, 11].

The discovery of the amorphous silicon solar cells caused excitement because it was a fundamentally cheaper material than standard crystalline silicon. This material is also used for the thin film transistors that drive modern flat panel displays. They have been massively produced and cheaply, and over the years the price for flat panels has decreased significantly. The same is happening for solar cell prices.

In the 1980s the solar PV community made steady progress toward improving efficiency, and many new types of solar cells were introduced. So, by the 1990s there was large-scale production of solar cells that were more than 10 % efficient, using the following materials: Ga-As and other elements of groups III and V of the periodic table, CdTe and $CuInSe_2$, TiO_2 dye sensitized, and finally the largest producers crystalline, polycrystalline, and amorphous silicon solar cells. Today prices continue to drop, and new "third-generation" solar cells are being actively researched upon [12–18].

The Energy's National Renewable Energy Laboratory (NREL) has set a world record in solar cell efficiency, with a PV device that converts 40.8 % (August 2008) of the light that hits it into electricity. This could be the highest confirmed efficiency of any such device to date [19].

The prospective energy in a PV power generation system is unlimited, and hence its use is environmentally friendly and is possible as a long-term reliable source of energy. PV systems can be used as an option to achieve renewable energy goals and to meet the goal of reducing emissions of carbon dioxide (Fig. 6.5).

Recently, globally installed PV systems have rapidly increased, with a growth rate of 8 years in a row of about 30 %. Total installed capacity of PV installations in the world reached 1242 MW in 2003, and the data in 2007 attained 62 % growth over that of the previous year, hitting a new height of 2826 MW. In 2007, the revenues from the PV industry worldwide reached 17.2 billion US dollars, of which Germany's PV reached 1328 MW, accounting for 47 % of the world market.

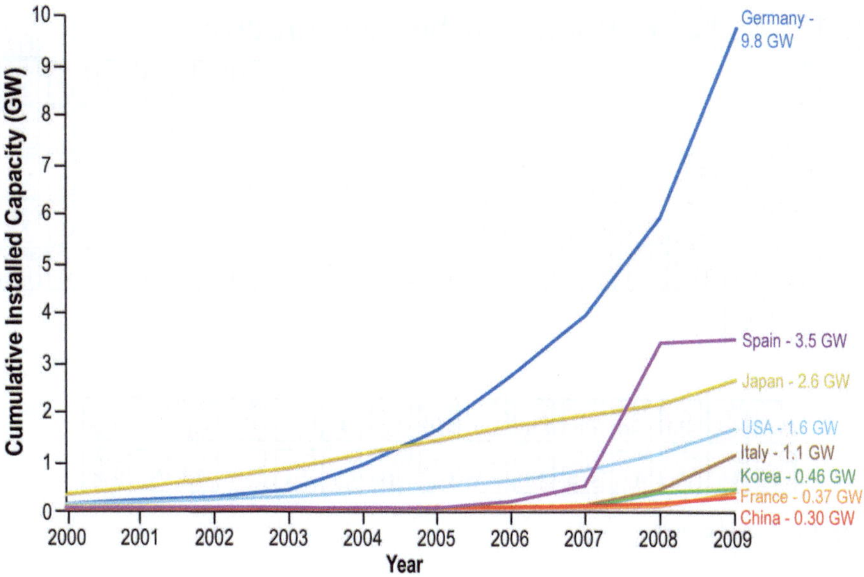

Fig. 6.5 PV cumulative installations [24]

Spain's PV market grew more than 48 %, reaching up to 640 MW. The USA increased by 57 %, to more than 220 MW, accounting for nearly 8 % of the world market, and became number 4 major market in the world behind Japan. These four countries accounted for 86 % of the solar PV market [20–23].

6.6 Photovoltaic Materials

Solar cells are made from semiconductors—the most important one being silicon. Semiconductors have special electronic properties which allow them to be conducting or insulating depending on their composition. In PV materials, you are dealing with the semiconductors (yellow). Most of the doping comes from phosphorous and boron for silicon solar cell sate. There are different types of solar cells such as cadmium telluride (CdTe) made from cadmium and tellurium or gallium arsenide or copper indium diselenide (Fig. 6.6).

Silicon is semiconductor from group 4 elemental. Silicon crystal forms the diamond lattice resulting in the use of four valence electrons of each silicon atom as shown in Fig. 6.7. Most solar cells are made from silicon. Each silicon atom in the lattice is in an identical position to every other atom, and each has four nearest neighbors. The four electrons in Si form bonds to the nearest neighbors (as seen by the red bonds in Fig. 6.8). Silicon is a semiconductor and is the element largely responsible for the integrated circuit, which makes modern computers possible [24].

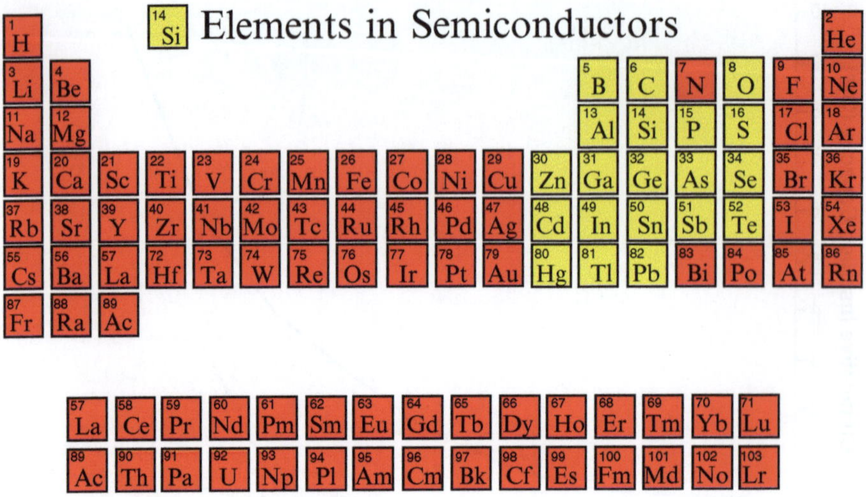

Fig. 6.6 Periodic table

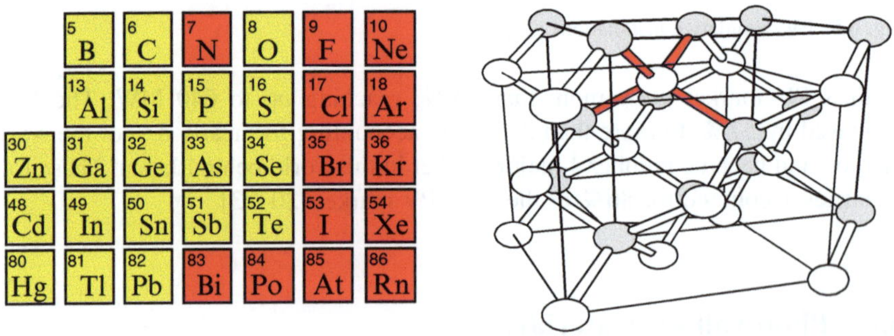

Fig. 6.7 Silicon crystal

Let's take a look at silicon technology first because it makes up the majority of the current PV market. Here is the graphical difference between crystalline silicon and amorphous silicon. The crystalline silicon lattice is perfect like the swan. The amorphous is messed up with defects (strained and broken bonds) like the ugly duckling. These physical defects lead to electronic defects and lower solar cell performances.

6.6.1 Band Theory

In band theory, there are three types of materials which are differentiated by their electronic structure: insulators, conductors, and semiconductors. In fact, band theory models the behavior of electrons in solids by postulating the existence of

Fig. 6.8 Comparison between crystalline and amorphous silicon [4]

energy bands, continuous ranges of energy which electrons may occupy, and gaps, which they may not. It successfully explains many physical properties of solids, such as optical absorption and electric resistivity. For the conductor, the valence band (E_v) is above conduction, so there are free electrons to conduct electricity. In the insulator, the two bands are far apart from each other, and it's very hard to get charge carried from the bottom (valence) to top band (conduction band E_c) [1]. For the semiconductor, there is a gap between the valence and conduction bands—this is known as the bandgap and is denoted as E_g *(energy of the gap)*. The *Fermi energy* *(E_f)* is also shown for the insulator and semiconductor. The ground state of a noninteracting fermions system (like one made up of electrons here) is constructed by starting with an empty system and adding particles one at a time, in a row filling up the lowest-energy unoccupied quantum states. The Fermi energy is the level (if there is one) where half of the states are occupied with electrons.

6.6.2 Energy Bands in a Semiconductor

There are three types of semiconductors as shown in Fig. 6.9:

1. Intrinsic
2. n-type
3. p-type

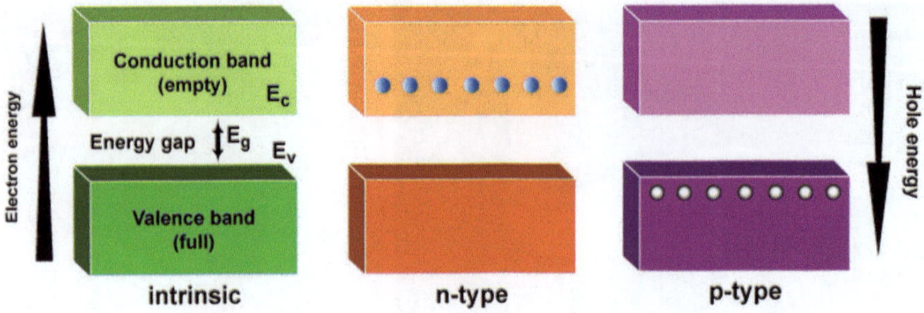

Fig. 6.9 Intrinsic, n-type, and p-type semiconductor

The semiconductors that conduct electricity are types 2 and 3:

- By alloying semiconductor with an impurity also known as doping
- Carriers located in conduction band or carriers removed from valence band

6.6.3 The Intrinsic Type

The intrinsic or pure semiconductor contains the right number of electrons to fill valence band. Consequently, the conduction band is empty. The pure semiconductor acts like an insulator because the electrons in full valence cannot move.

6.6.4 The n-Type

An atom from group 5 impurity added to silicon melt from which crystal is grown so that the current is carried by the negatively charged electrons.

- The valence band filled by 4/5 of outer electrons.
- 1/5 left is then put into conduction band. These impurity atoms are called "donors."

The electrons move within conduction band; therefore, crystal becomes a "conductor."

6.6.5 The p-Type

Group 3 added to silicon melt make current carried by missing electron holes to act as positively charged particles.

- Doping creates lack of electrons in the valence band which needs 4 out of 3 outer electrons.
- Missing electrons, also known as holes, are used to carry current.

The existing charges are called the majority carriers. The existing charge carriers in n-type are "electrons." The existing charge carriers in p-type are "holes."

6.7 Creating a Junction

The semiconductor junctions can be classified into four main types: p–n, p–i–n, Schottky barrier, and heterojunction. Each has a built-in potential. You can create devices by putting two similar semiconductors together. The device is called a junction and there are four different types. Each has a built-in potential (a built-in voltage) associated with them [6].

Here are the first two types of junctions, the p–n and p–i–n junctions. You have the conduction band and the valence band. No electrons can exist in bandgap; however, semiconductors are not perfect. In thermodynamic equilibrium at room temperature, you actually populate some of the states in this bandgap. The solid line is the Fermi level. You can think of it as electrons as a liquid being poured into a bucket (the gap) and being filled up to the Fermi level. In semiconductors, when you put two dissimilar semiconductors together, the Fermi level must stay flat. This is what created the bending of these bands (both the conduction and the valence band).

So, in the p–n junction you have the p-layer and n-layer. In the p-type, the Fermi level is very low in the gap pulled to the valence band. In the n-type semiconductor, the Fermi level is very high and pulled toward the conduction band. There is an enormous bending of the conduction and valence bands due to trying to keep Fermi level flat when they are combined. The bending creates the voltage and the electric field observed when putting two dissimilar (p and n) semiconductors together. The energy between the conduction band in the p-layer and the n-layer is the built-in potential (Fig. 6.10).

All the junctions contain strong electric field. The electric field occurs as follows:

- When two semiconductors approach into contact, from n-type, the electrons near the interface transfer over to p-type and send off a positively charged area.
- By the interface, holes from p-type transfer over to n-type and send off a negatively charged area.
- Because holes and electrons are swapped, a middle potential barrier with no mobile charges is formed.
- This potentially created barrier does not let any more electrons or holes flow through.
- The electric field pulls holes and electrons in opposite directions, which means there is no net current and equilibrium occurs.

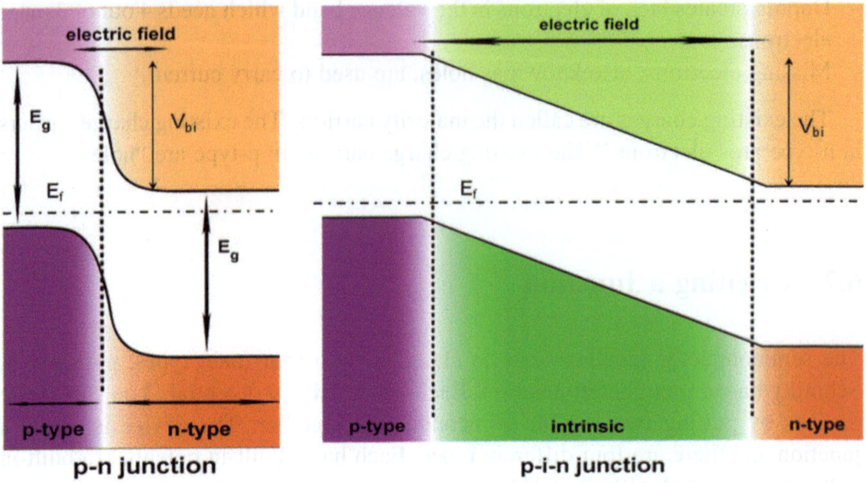

Fig. 6.10 p–n and p–i-n junctions

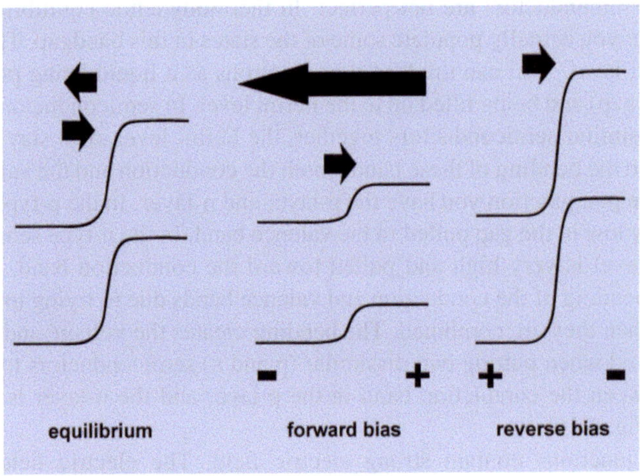

Fig. 6.11 Barriers charges

- The positive voltage applied to p-side leads to reduced barrier height, and it is called "forward bias," which results in increases of the current through PV.
- Increased barrier height is called "reverse bias," which results in decreases of the current to a very small amount (Fig. 6.11).

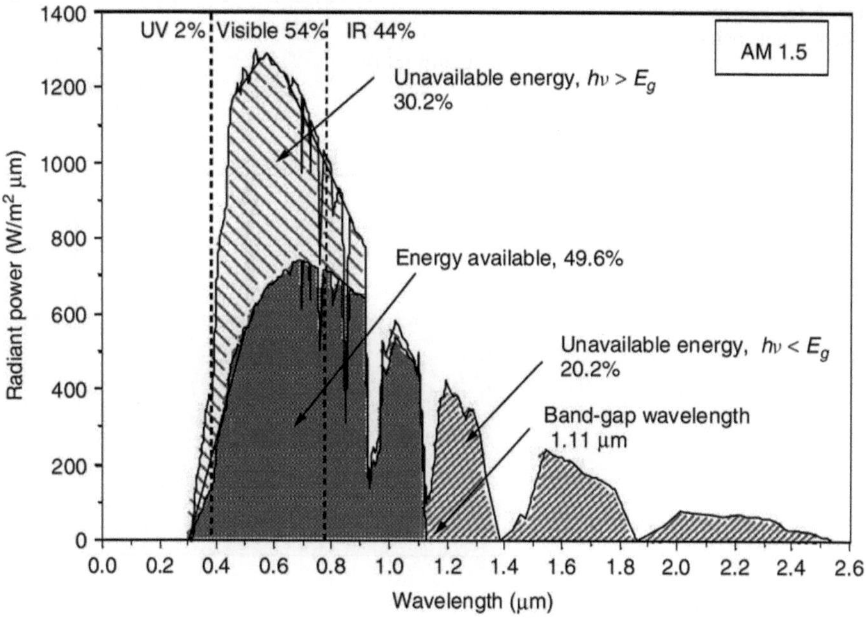

Fig. 6.12 Solar spectrum [25]

6.8 The Solar Spectrum and Generation of Electron Hole Pairs with Light

Figure 6.12 shows the solar spectrum. The graph shows the flux of photons that come down to the earth vs. the wavelength of the photons. The orange line represents the solar spectrum outside the atmosphere. The spectrum reaches a peak in the visible as expected from black-body radiation at 6000 K (the surface temperature of the sun). The missing regions of flux (black line) at the earth's surface are due to absorption from water vapor and carbon dioxide [7]. As you can clearly see, the solar energy is not just a single energy (wavelength or frequency of light). If it were, we would have solar cells based on current technology with efficiencies over 95 %. So, why do we lose energy at different wavelengths? Part of it is because we are not absorbing the whole spectrum of a single photon. If we put in photons with larger energies than the bandgap, we only capture the energy equal to E_g—we lose the rest to waste heat. Photons with energies lower than the bandgap do not contribute to carriers (electrons and holes) in a solar cell.

Photon is absorbed when it enters and lets electron from V_B get sent up to C_B. Consequently, a hole is left behind in V_B, creating absorption process: electron-hole pairs. Since only part of solar spectrum can be converted, the photon flux transformed by a solar cell is about 2/3 of total flux (Fig. 6.13).

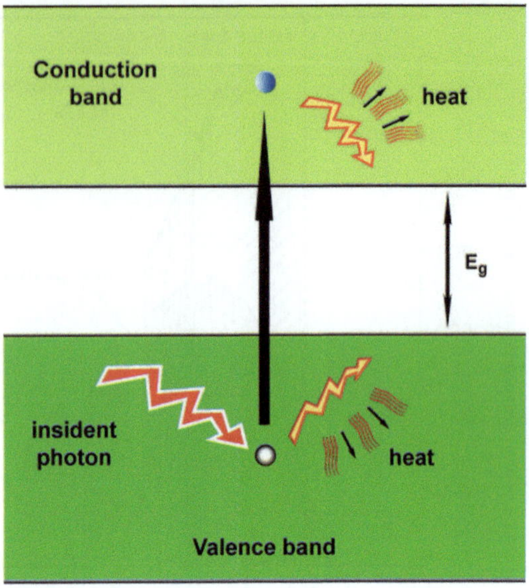

Fig. 6.13 Generation of electron-hole pairs

6.9 Electric Currents in p–n Junction Under External Bias

Figure 6.14 shows a typical current (*I*) and voltage (*V*) plot for a p–n junction known as *I–V* characteristics (or curve) [6], applying a voltage from negative to positive and recording the current. For the dark current, it is not quite zero, and it starts to go up and pushes electrons over creating a massive current. When illuminated, you instantly end up with a current [1]. The power is calculated as the current multiplied by the voltage ($P = IV$). We will talk about extracting power from a semiconductor p–n junction in a moment.

The output current of p–n junction is

$$I = I_L - I_o\left[e^{qV/kT} - 1\right] \tag{6.30}$$

- I_L = light generated current.
- q = electric charge.
- V = voltage.
- k = Boltzmann's constant = $1.3807 \times 1^{-23}\ J/K$.

  - All light generated current passes through the diode (or PV) when in open circuit ($I = 0$).
  - All current passes through external load when in short circuit ($V = 0$).

Two important points have to be mentioned here:

1. The voltage of open circuit during open-circuit state

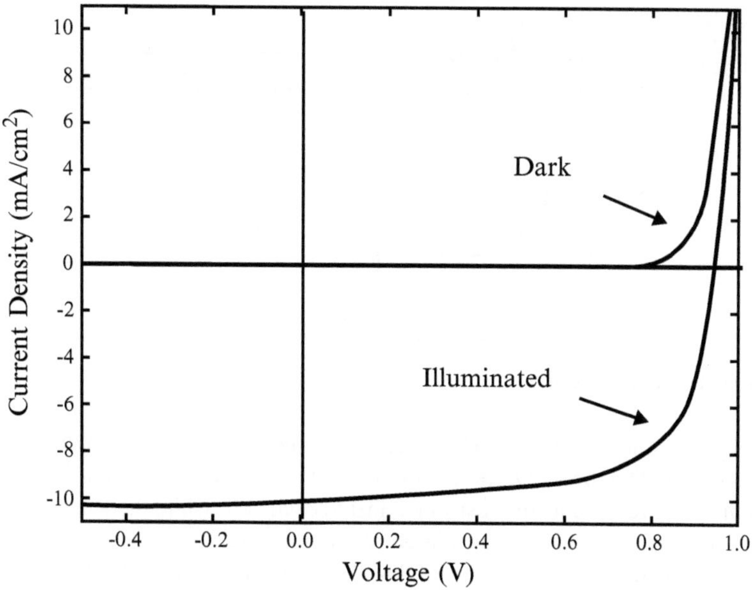

Fig. 6.14 Typical I–V characteristics

$$V_{oc} = \frac{kT}{q} \ln\left(\frac{I_L}{I_o} + 1\right) \tag{6.31}$$

2. Under short and open circuit, no power is generated but

$$P_{max} = V_{max} I_{max} = FF.V_{oc} I_{sc} \tag{6.32}$$

The fill factor (*FF*) is a measure of the real *I–V* characteristic. For efficient solar cells (p–n junction), the value should be greater than 0.7. The fill factor diminishes as the cell temperature increases.

6.10 Solar Cell Modeling

The simplest model or equivalent circuit of a solar cell is a current source connected in parallel with a p–n junction (diode) as shown in Fig. 6.15. The output of the current source is directly proportional to the light falling on the cell. During darkness, the solar cell is not an active device and it works as a diode. It produces

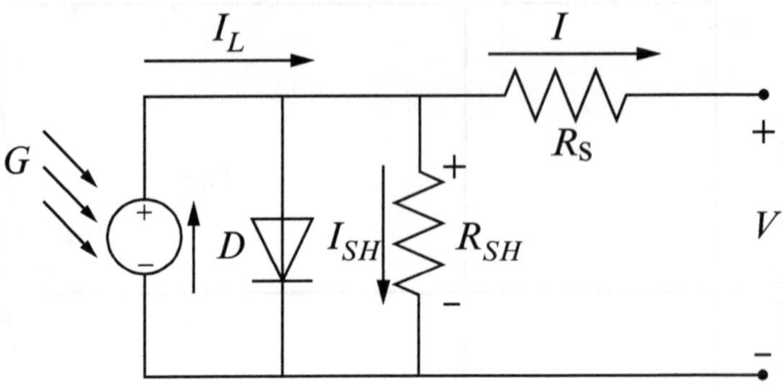

Fig. 6.15 Equivalent circuit of solar cell

neither current nor voltage [5]. However, if light falls on the solar cell, it generates a diode current. The diode, D, determines the I–V characteristics of the cell. A series resistance, R_S, represents the resistance inside each cell, while the shunt resistance, R_{SH}, is neglected because it has a large resistance value.

In an ideal solar cell, it is assumed that $R_S = 0$ and $R_{SH} =$ infinity. The net current of the cell could be found by improving Eq. 6.30 as follows:

$$I = I_L - I_o\left[e^{q(V+IR_S)/nkT} - 1\right] \tag{6.33}$$

The photocurrent, I_L, depends on reference first and second temperatures, T_1 and T_2, respectively, and it is given by

$$I_L = I_L(T_1) + K_o(T - T_1) \tag{6.34}$$

where

$$I_L(T_1) = I_{SCT1,nom}(G/G_{nom}) \tag{6.35}$$

$$K_o = (I_{SCT2} - I_{SCT1})/(T_2 - T_1) \tag{6.36}$$

where G is the present solar radiation and G_{nom} is the solar radiation at the reference test.

The saturation current of the diode, I_o, is given by

$$I_o = I_{oT1}(T/T_1)^{3/n} e^{\frac{qV_{qT1}}{nk\left(\frac{1}{T}-\frac{1}{T_1}\right)}} \tag{6.37}$$

where

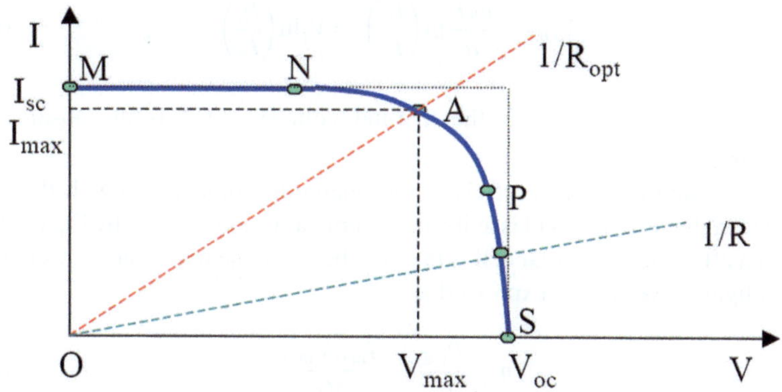

Fig. 6.16 I–V characteristic curve of a solar cell

$$I_{oT1} = I_{SCT_1} / e^{\frac{qV_{qT1}}{nk\left(\frac{1}{T}-\frac{1}{T_1}\right)}} \tag{6.38}$$

The series resistance of a solar cell is given by

$$R_S = -\frac{dV}{dI_{VOC}} - \frac{1}{X_V} \tag{6.39}$$

where

$$X_V = I_{oT_1} \frac{q}{nkT_1} e^{\frac{qV_{OCT1}}{nkT_1}} \tag{6.40}$$

A typical *I–V* characteristic of a solar cell at a certain ambient irradiation, *G*, and fixed cell temperature is shown in Fig. 6.16 [6]. The load characteristic is a straight line for a resistive load with slope *I/V* = *1/R*. It is noted that the power delivered to the load is proportional to the value of resistance.

From Fig. 6.16, the cell operates in the region M–N of the curve, if the load is small, where the cell behaves as a constant current source which is almost equal to the short-circuit current. On the other hand, the cell operates in the regions P–S of the curve, if the load is large and the cell behaves more as a constant voltage source which is approximately equal to the open-circuit voltage.

The short-circuit current, I_{sc}, is the maximum current value generated by a PV or solar cell. It is produced during the short-circuit condition where *V* = *0*. The open-circuit voltage V_{oc} corresponds to the voltage drop across the diode when the photocurrent is zero. It reflects the voltage of the cell at no light conditions, and it can be mathematically expressed as

$$V_{OC} = \frac{nkT}{q}\ln\left(\frac{I_L}{I_o}\right) = V_t\ln\left(\frac{I_L}{I_o}\right) \tag{6.41}$$

where $V_t = \frac{mkT_C}{q}$ is known as the thermal voltage and T is the absolute cell temperature.

The "maximum power point" is at the operating point A in which the power dissipated in the resistive load is at its maximum value and is given by Eq. 6.32. The maximum efficiency of a solar cell is the ratio between the maximum power and the incident light power and is expressed as

$$\eta = \frac{P_{\max}}{P_{in}} = \frac{I_{\max}V_{\max}}{AG_a} \tag{6.42}$$

where A is the area of the PV module and G_a is radiation.

6.11 The Solar Cell, Module, and Array

In semiconductor the electric current generated is extracted by contacts to the front and rear of cell. Fingers, which are the extensively spaced thin strips, are created so that light is permitted through. These fingers supply current to the larger bus bar. Antireflection coating (ARC) is used to cover the cell to minimize light reflection from the top surface. ARC is made with thin layer of dielectric material [24] (Fig. 6.17).

The one-half volt produced by a single solar cell is inadequate for almost all applications. Consequently, to increase the voltage, cells are connected together in series. Several of these series strings of cells may be connected together in parallel to increase the current as well. These interconnected cells and their electrical connections are then sandwiched between a top layer of glass or clear plastic and a lower level of plastic metal. To increase mechanical strength, an outer frame is attached to provide a way to mount the unit. This package is called as a PV panel or PV module. The power available from one module is insufficient for supplying power to many loads. A number of PV modules can be connected in parallel, series, or both so as to form a PV array to increase either output current or voltage.

6.12 Photovoltaic Systems

Photovoltaic systems are generally classified according to operational requirements and their functional, component configuration and how the equipment is connected to the other power sources and electrical loads. The two classifications of PV systems are stand-alone and grid-connected systems. The stand-alone PV systems

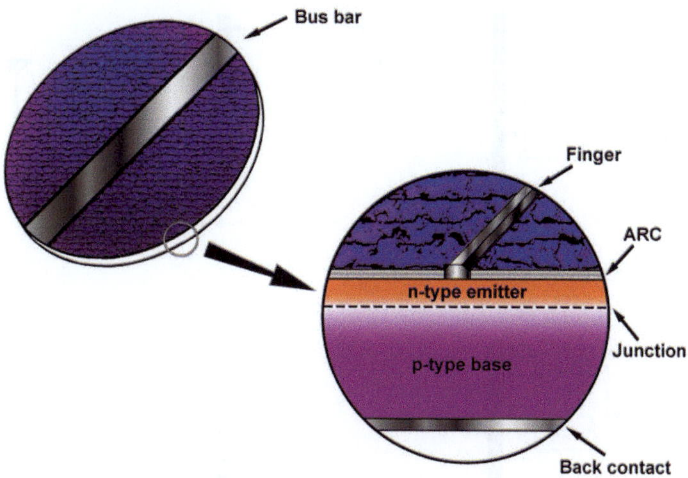

Fig. 6.17 Solar cell

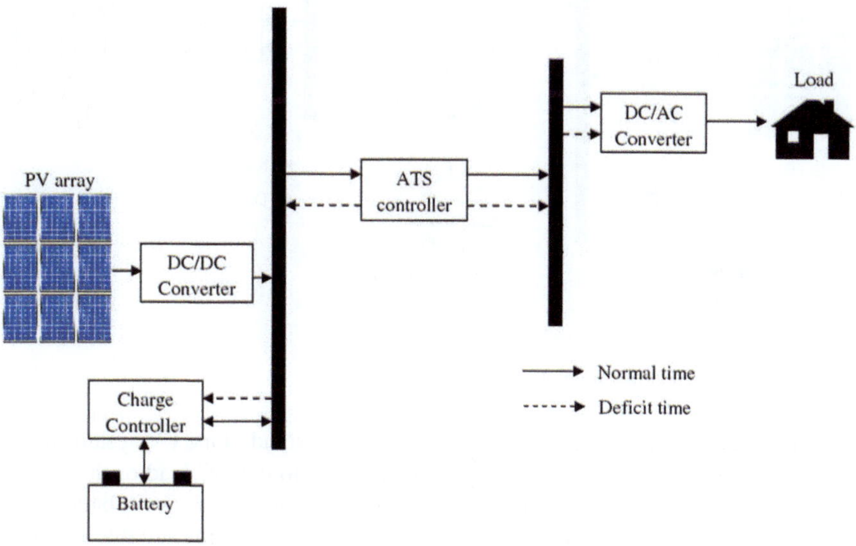

Fig. 6.18 Stand-alone PV system

[3] as shown in Fig. 6.18 are considered to operate independent of the electric utility grid [26] and are normally sized and designed to supply certain DC and/or AC electrical loads. Stand-alone systems may be powered by only a PV array as shown in Fig. 6.19. If an engine generator, wind generator, or utility power is used as a backup power source, the system is called as a PV hybrid system as shown in Fig. 6.20 [27].

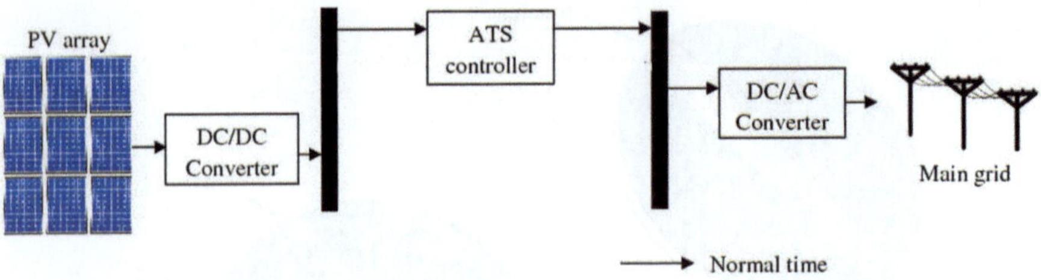

Fig. 6.19 Grid-connected PV system

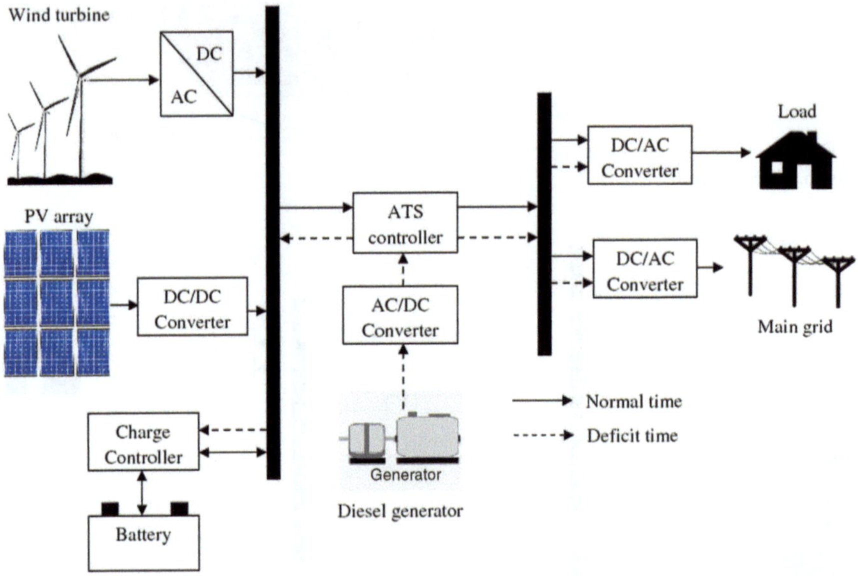

Fig. 6.20 Hybrid PV/wind/diesel generator system

The direct-coupled system is the simplest type of stand-alone PV system where the PV array output terminals are connected directly to the DC load or through an inverter to the AC load. In the direct couple system, since there is no battery, the load only operates during sunshine hours, making these designs suitable for common applications such as water pumps, ventilation fans, and small circulation pumps for solar thermal water heating systems. Matching electrical load impedance to the PV maximum output power is a significant issue in designing the direct-coupled system. DC to DC converter, which is known as maximum power point tracker (MPPT), is used in order to better utilize the PV array maximum output power. Most of PV systems use batteries for energy storage.

The power generated from PV modules depends fundamentally on the amount of solar energy captured by the module where the amount of captured energy depends on the angle at which the PV module has been installed. Thus, if a PV array tracks the sun, the energy yield increases. In Europe, relatively high radiation gains can be obtained by means of a tracking mechanism, on days with high irradiation and large proportion on direct radiation. These gains can reach about 50 % on clear days in summer and in winter 300 % as compared with systems with a static horizontal PV array [28]. A major part of the increase in yield due to sun tracking can be obtained in summer with absolute gains in comparison with winter.

6.13 Photovoltaic Systems Design and Optimization

In rural, isolated, and desert area, the electric provisioning of electrical sector is done by the solar photovoltaic (PV) or hybrid systems of production of electricity whose diesel and natural gas generators play a significant role like auxiliary source. These hybrid systems engage combination of different energy sources like PV/battery, wind/battery, wind/PV/diesel/battery, wind/ PV/battery, etc.

However, the designed PV/battery system must meet the desired electrical load at a definite level of security. In literature lots of optimization methods for PV/battery system can be found. Based on the reviewed work, it is found that there are three main systems design and sizing procedures, namely, intuitive, numerical (simulation based), and analytical techniques in addition to some individual methods as shown in Fig. 6.21 [29].

6.13.1 Intuitive Technique

The intuitive technique is defined as "a simplified calculation of the size of the system carried out without establishing any relationship between the different subsystems or taking into account the random nature of solar radiation."

Firstly in designing the PV power system, an estimation of the energy demand of the load should be done by multiplying the power of each appliance by the average number of hours of use. Then 20 % might be added to allow for losses caused by wiring [28], DC to AC conversion, dirty modules, etc. Loads, whatever AC or DC loads, should be described in a worksheet by load current, load voltage, daily duty cycle, weekly duty cycle, power conversion efficiency, nominal systems voltage, and amp-hour load. The designer should consider energy-conserving substitutes for items that are used often. Identify large and/or variable loads, and determine if they can be eliminated or changed to operate from another power source. LED lamps should be used in place of incandescent lamps. They provide better light levels with much lower power demand. The operating voltage selected for a stand-alone PV system depends on the total current and the voltage requirements of the loads. If the system voltage is located equally to the largest load voltage, then these loads may be connected directly to the system output, though it is recommended that the current in any source circuit be kept below 20 with a 100 A limit for any section of the system. Keeping the current below, these recommended levels will allow the use of standard and commonly available electrical hardware and wires. When loads

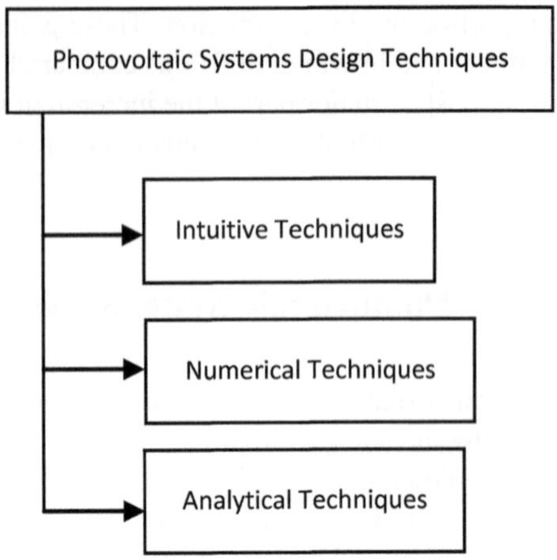

Fig. 6.21 PV system design techniques

require AC power, the DC system voltage should be selected after studying available inverter characteristics [3].

The optimum selection of the inverter represented by choosing an inverter meets the load and keeps the DC current below 100A. The selection of an inverter is important and affects both the cost and performance of the system. Generally, power-handling capability and efficiency are better for units operating at higher DC voltages; a 48 V unit is usually more efficient than a 12 V unit. The designer should obtain information on specific inverters and their availability, cost, and capabilities, from several manufacturers before making the decision on system voltage. It is worth mentioning to consider the basic building block in the storage subsystems, and PV array gets larger as the voltage increases. For example, a 24 V solar system has two series-connected PV modules to form the basic building block though the higher operating voltage advantage is the lower current required to produce the same power. However, high current leads to large wire size, which means an increase in the expensive and hard-to-get switches, connectors, and fuses. Finally, prior information of the components' availability and cost and switchgear is significant to high-quality system design [4].

For stand-alone system shown in Fig. 6.22, the required PV modules and battery capacity can be calculated using some of the formulas below:

$$P_{PV} = \frac{E_L}{\eta_S \eta_{inv} PSH} S_f \tag{6.43}$$

where E_L is the daily energy consumption, PSH is the peak sun hours, η_{inv} and η_S are the efficiencies of the system components, and S_f is the safety factor that represents PV cell temperature losses and the compensation of resistive losses [30].

On the other hand, the battery capacity can be calculated by

$$C_{Wh} = \frac{E_L \times D_{Autonomus}}{V_B DOD \eta_B} \tag{6.44}$$

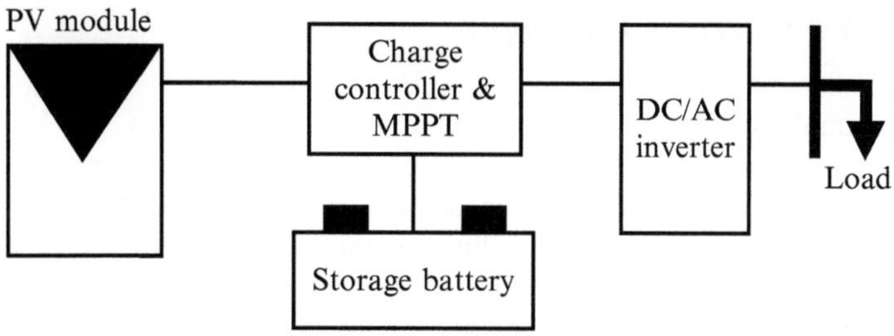

Fig. 6.22 Typical PV system components

where η_B and V_B are the efficiency and voltage of the battery block, respectively, while *DOD* is the permissible depth of discharge rate of a cell. However, the PV array energy can be expressed in terms of ambient temperature and solar energy as

$$E_{PV}(t) = A_{PV}E_{sun}(t)\eta_{wire}\eta_{PV}\eta_{inv} \tag{6.45}$$

where A_{PV} and E_{sun} are the area of PV array and solar energy, respectively. The efficiency of wire, PV conversion and inverter conversion, is η_{wire} and η_{PV}, respectively. However, the conversion efficiency of a PV array is effected temperature and expressed as follows:

$$\eta_{PV}(t) = \eta_{PV_{\mathrm{Ref}}}\left[1 - \beta_T\left(T_c(t) - T_{C_{\mathrm{Ref}}}\right)\right] \tag{6.46}$$

where η_{pvRef} and β_T are the reference PV module conversion and temperature coefficient efficiencies, respectively. T_{cref} and T_c are the reference cell and cell temperatures, respectively. The ambient temperature is used to calculate cell temperature as follows:

$$T_c(t) - T_{ambient} = \frac{T_{Test}}{800}G(t) \tag{6.47}$$

where T_{Test}, $T_{ambient}$, and G are the temperature during testing, the ambient temperature, and the PV module and solar radiation, respectively.

The annual capacity factor is defined as "the ratio of actual annual energy output to the amount of energy the PV array would generate if it operates at full rated power (P_r) for *24* h per day for a year":

$$CF = YF/8760 = E_{PVannual}/(P_R \times 8760) \tag{6.48}$$

Meanwhile, the yield factor is defined as "the daily, monthly, or annual net AC energy output of the system divided by the peak power of the installed PV array":

$$YF = E_{PV}(kWh/year)/PV_{WP}(kW_p) \tag{6.49}$$

6.13.2 A Numerical Technique

A system simulation is used in the case of a numerical technique. The battery load state and energy balance of the system are calculated for each time period considered, usually an hour or a day. The advantages of these techniques are their accuracy comparing with intuitive technique, and the thought of energy reliability can be applied in a quantitative manner. There are many programs for PV design and modeling used to estimate the power output of a PV system, which typically includes different components: PV panels, charge controllers, inverters, etc. Some of these programs are PVSS, Evans and Facinelli Model, SOLCEL, PVSIM, PVForm, Sandia Inverter Performance Model, Sandia Photovoltaic Array Performance Model, Solar Advisor Model, PV-DesignPro, Five-Parameter Array Performance Model, PVSYST, PVWatts, RETScreen Photovoltaic Project Model, PV F-Chart, Polysun, PVSol, SolarPro, INSEL, Clean Power Estimator, PVOptimize, OnGrid, CPF Tools, Solar Estimate, SOLSTOR, HybSim, Hysim, Hybrid2, UW-Hybrid (TRNSYS), RETScreen, PVToolbox, RAPSIM, SOMES, IPSYS, HySys, Dymola/Modelica, HOMER, PV.MY, and REPS.OM [31, 32].

6.13.3 An Analytical Technique

In the analytical technique, the size of the PV/battery system described by equation as a function of the reliability is to develop. The major advantage of this technique is that the design and the system size calculation are very straightforward, while the disadvantage of this technique is that these equations are location-dependent factors as well as the difficulty of finding the coefficients of these equations is another point.

6.14 PV Systems Economic

The expressions and terms related to the economics of renewable energy systems are presented in this section. An example will be given on PV system, but the concept is the same for other renewable energy systems. The main significant PV economic parameters are the capital cost (PV system total installing cost), feed-in tariffs, the energy payback time (EPBT), and electricity price. Investments into renewable energies, particularly into PV technologies, are other economics-related areas [2]. The PV system cost is usually measured by price per peak watt (e.g., USD/Wp or €/Wp). Peak watt is defined as "the power at Standard Test Conditions STC (solar irradiance 1000 W/m2, AM of 1.5 and temperature 25 °C)." PV system costs encompass both module and BOS costs. The PV module costs normally represent simply 40–60 % of total PV system costs. Around half of this investment would be for the PV modules/array and the inverter, electrical cabling, PV array support structures, and the rest accounted for equipment and installation. The BOS and installation costs can vary significantly. For example, the total installation costs are high when costs for site preparation, system design and engineering, laying a foundation, permitting, as well as installation labor and assembly are high [4]. The life cycle cost (*LCC*) of a PV system may also include costs for system design and engineering, site preparation, installation labor, permits, and operation and maintenance costs. PV systems have an anticipated 20–25-year lifetime [27].

The different components taken into account for calculating *LCC* are capital cost, replacement cost, maintenance cost, and salvage value.

The capital cost is the initial cost that is spent for the purchase of PV modules, batteries, charge controller, inverter, and mechanical structures to support PV modules, circuit breakers, special cables, and installation charges. The replacement cost is the cost incurred for replacing batteries, charge controller, inverter, and the cables at the end of their lifetime. The batteries have to be replaced once every 6 years and the charge controller, inverter, and the cables once in 12 years. The maintenance cost is the cost incurred every year as annual maintenance cost. The salvage value is the money that is obtained while disposing the solar modules at the end of life cycle period. For calculating *LCC*, each of the costs is converted to its present worth [3]. Even the salvage amount is converted to its present worth for calculating *LCC*.

Payback period and cost of energy are the criteria that are used to moderate the system economically. About 40 % of the PV system investment is for the PV modules, while the rest accounts for support structure, inverter, cabling, equipment, and installation. The *LCC* of a PV system may also include costs for site preparation, system design, installation labor, permits, and operation and maintenance costs [30]. The life cycle cost is calculated using the following equation:

$$LCC = C_{capital} + \sum_{1}^{n} C_{O\&M}.R_{PW} + \sum_{1}^{n} C_{replacement}.R_{PW} - C_{salvage}.R_{PW} \quad (6.50)$$

The different components taken into account for calculating *LCC* are capital cost $C_{capital}$, replacement cost $C_{replacement}$, maintenance cost $C_O\&_M$, and salvage value $C_{salvage}$. RPW represents the present worth of each factor which is calculated using the future sum of money (F) in a given year (N) at a given discount rate (I):

$$R_{PW} = F/(1+I)^N \quad (6.51)$$

After calculating *LCC*, the cost of energy is calculated using [26]

$$CoE = LCC/\sum_{1}^{n} E_{PVannual} \quad (6.52)$$

where $E_{PVannual}$ is the annual energy production of the PV system, while n is the system lifetime in years.

Finally, the payback period is calculated as follows:

$$PBP = C_{capital}(USD)/[E_{PVannual}(kWh/year) \times CoE(USD/kWh) \times R_{PW}] \quad (6.53)$$

6.15 Photovoltaic Applications

The PV systems' gifted source of energy is used to meet the energy need, directly converting sunlight into electricity with no harmful pollution and no moving parts, even though more conventional energy sources, such as fossil fuels (i.e., oil, gas, etc.), are still providing most of the global energy demand. The PV systems have a variety of applications ranging between small, medium, and large scales. The PV applications can be classified into three categories: stand-alone, utility interactive, and solar tracking systems.

The utility interactive systems are particularly striking in displacing electrical power bought from the utility during maximum demand hours, which usually coincide with maximum sunlight hours. The peak demand surcharge included in the commercial customer's bill is electrical companies so that PV systems can significantly reduce these bills [2].

The stand-alone systems' PV technology and systems have a wide range of applications. A stand-alone system is one that does not connect to the utility. Some examples of the various applications of stand-alone PV systems include:

Lighting: The PV systems used to supply lighting for highway information signs, billboards, parking lots, public-use facilities, homes, marinas, piers, trains caboose lighting, and vacation cabins.

Communications: The PV systems used to supply television, radio, and phone signals over long distances need to be amplified. The repeater stations or relay towers perform this function. The best location for relay towers is typically at the maximum potential height where utility power lines are not commonly found and the transport of conventional generator fuels would be costly and difficult. PV also is used on portable computer systems, travelers' information transmitters, mobile radio systems, cellular telephones, and emergency call boxes.

Remote Site Electrification: Numerous residential loads and other structures which are far to connect with the utility grid. There are several examples of remote site electrification such as visitor centers in parks, rural homes, vacation cabins, and park ranger sites, remote farm workshops, hunting lodges, clinics, village/island electrification, and remote research facilities, public beach facilities, highway rest stops, military test areas and campgrounds.

Remote Monitoring: For scientific research regularly monitoring or other purposes temporarily take place at sites far from utility connections. PV systems can be efficiently used as a power supply to weather station, monitoring of meteorological information, structural conditions, highway/traffic conditions, irrigation control, seismic recording, and scientific research in remote locations.

Signs and Signals: Devices and equipment as example audible warning signs such as sirens, navigational beacons, railroad signals, highway warning signs, and aircraft warning beacons, buoys, and lighthouses are unfeasible to connect to the utility and generally remote. PV systems present reliable power supply for these significant applications.

Water Pumping and Control: PV is an ideal applicant for water pumping applications. Water pumping systems supplied by PV technologies are used to provide water for irrigation, campgrounds, livestock watering, and water supplies of remote village.

Charging Vehicle Batteries: Directly charge vehicle batteries can use PV systems as power supply. Also, PV could provide a "trickle charge," on little-used vehicles, for maintaining a high battery state of charge, such as snow removal equipment, and fire-fighting and agricultural machines such as harvesters or tractors. Direct charging is useful for recreational vehicles and boats.

Disaster Relief Applications: Natural disasters such as floods, hurricanes, earthquakes, and tornadoes often destroy electric generating facilities and utility systems. When there is electrical power blackout for a period of time which may affect cities and areas, portable PV systems are a very good option for providing personal and street lighting, warning and message signs, power for communications equipment, medical supplies refrigeration, water purification, and pumping water.

Cathodic Protection: When metallic equipment is exposed to water and soils obviously experience corrosion because of electrolytic action. Metals lose ions which lead to this corrosion when metal is exposed to an electrolyte. To avoid ion loss from metal, a small voltage could be applied to prevent corrosion. Cathodic protection is the name of this method. Only a small DC voltage is essential to protect the metals. To use utility as power supply, the voltage must be stepped down and converted from AC to DC. PV systems and technologies are able to produce the needed DC voltage for cathodic protection simply, efficiently, and cheaply. Cathodic protection is used on tanks, pipes, wharves, wellheads, buildings, and bridges.

Refrigeration: PV systems present good option for mobile or remote storage of medicines and vaccines.

Consumer Products: PV systems are used for different applications which are available commercially and used by consumers. Examples of PV applications include lanterns, watches, radios, calculators, recreational vehicles, flashlights, televisions, security systems, outdoor lights, golf carts, fans, and gate openers.

References

1. Aldo V. DaRose, *Fundamentals of renewable energy processes*, (Elsevier, 2005)
2. Hussein A. Kazem, *Renewable and sustainable energy, principles and applications*, 1st edition, 978–3-659-46238-2 (LAMBERT Academic Publishing, Germany, 2013)
3. H.A. Kazem, T. Khatib, K. Sopian, Sizing of a standalone photovoltaic/battery system at minimum cost for remote housing electrification in Sohar, Oman. Energy Build. **6C**, 108–115 (2013)
4. H.A. Kazem, T. Khatib, Techno-economical assessment of grid connected photovoltaic power systems productivity in Oman. Sustain. Energy Technol. Assess. **3**, 61–64 (2013)
5. Hussein A. Kazem, Tamer Khatib, *Photovoltaic power system prospective in Oman, Technical and economic study*, 1st edition, ISBN: 978-3659372957, (LAMBERT Academic Publishing, Germany)
6. Tomas Markvart, *Solar electricity*, 2nd edition, (Wiley, University of Southampton, UK)
7. V. Quaschning, *Understanding renewable energy systems* (Carl Hanser Verlag GmbH, Munich, Germany, 2005)
8. Alfred Smee, *Elements of electro-biology: or the voltaic mechanism of man; of electro-pathology, especially of the nervous system; and of electro-therapeutics*, (Longman, Brown, Green, and Longmans, London, 1849), p. 15
9. Nobel Price official, The Nobel prize in physics 1921: Albert Einstein. Nobel Price official, www.nobelprize.org
10. D.E. Carlson, C.R. Wronski, Amorphous silicon solar cells. Appl. Phys. Lett. **28**, 671–673 (1976)
11. R.J. Handy, Theoretical analysis of the series resistance of a solar cell. Solid State Electron. **10**, 765 (1967)
12. Allen Barnett, et al., Milestones toward 50% efficient solar cell modules, 22nd Photovoltaic European Solar Energy Conference, Milan–Italy, 3 Sept, 2007
13. European Directory of Renewable Energy, 1991
14. Godfrey Boyle, *Energy systems and sustainability: Power for a sustainable future*. (Open University, Oxford U Press, 2004)
15. R. Hill, Prospects for photovoltaics. Energy World, **208**, 8–11, original data updated by Hynes K. and Hill R. in 1999
16. IEEE Standard 1262–1995, *Recommended practice for qualification of Photovoltaic (PV) modules*, (IEEE, Piscataway, NJ, 1995)
17. W. Xiao, N. Ozog, W.G. Dunford, Topology study of photovoltaic interface for maximum power point tracking. IEEE Trans. Ind. Electron. **54**(3), 1696–1704 (2007)
18. Hussein Kazem, Bahiya Al-Rawahi, Galiya Al-Shabibi, Talal Al-Nabhani, Solar energy in sultanate of Oman, *International Renewable Energy Congress IREC2010*, Sousse Tunisia, 5–7 Nov, 2010, pp. 36–39
19. National Renewable Energy Laboratory. NREAL: NREL Solar cell sets world efficiency record at 40.8 percent, 21 Oct, 2008
20. Ting Chung Yu, Tang Shiuan Chien, Analysis and simulation of characteristics and maximum power point tracking for photovoltaic systems. *International Conference on Power Electronics and Drive Systems*, PEDS 2009, pp. 1339–1344
21. Photon International (February, 2009 and 2010)
22. Proceedings of the IEEE PVSC (US), the European PVSEC (EU), and the Pacific Rim PVSEC (Asia and Australia)
23. The US PV industry roadmap through 2030 and beyond. www.seia.org
24. Lawrence L. Kazmerski, Photovoltaic: history, technology, markets, manufacturing, applications, and outlook. *84th International Seminar in Brighton, Renewable Energy Policy, Security, Electricity, Sustainable Transport, Water Resources/Management and Environment*, Brighton, UK, 3–9 Jul 2011

25. Gilbert, Renewable and efficient electric power systems. (John Wiley & Sons,Hoboken, NJ, 2004)
26. H.A. Kazem, T. Khatib, K. Sopian, W. Elmenreich, Performance and feasibility assessment of a 1.4kW roof top grid-connected photovoltaic power system under desertic weather conditions. Energy Build. **82**, 123–129 (2014)
27. Hussein A. Kazem, Tamer Khatib, A novel numerical algorithm for optimal sizing of a photovoltaic/wind/diesel generator/battery microgrid using loss of load probability index. Int J Photoenergy, p. 8 (March 2013)
28. H. Mousazadeh, A. Keyhani, A. Javadi, H. Mobli, K. Abrinia, A. Sharifi, A review of principle and sun-tracking methods for maximizing solar systems output. Renew. Sustain. Energy Rev. **13**, 1800–1818 (2009)
29. Hussein A Kazem, Amira M. S. Al-Aisaee, K. Sopian, Ahmed Said Albusaidi, M.H. Albadi, Design and evaluation of photovoltaic systems using different techniques. International Conference on Advances in Electrical and Electronics Engineering (ICAEEE'2014), Kuala Lumpur, Malaysia, 15–16 Jan 2014
30. Hussein A. Kazem, Tamer Khatib, Ali A K Alwaeli, Optimization of photovoltaic modules title angle for Oman. 7th IEEE International Power Engineering and Optimization Conference PEOCO2013, Malaysia, 3–4 June 2013, pp. 700–704
31. Endecon Engineering, A guide to photovoltaic system design and installation, California Energy Commission, Version 1.0, June 14, 2001
32. Geoffrey T. Klise Joshua S. Stein, Models used to assess the performance of photovoltaic systems, Sandia National Laboratories, SANDIA REPORT, SAND2009-8258, Unlimited Release, Printed December 2009

Chapter 7
Photovoltaics and the Energy System: Adaptation of Layout and Load

Stefan Krauter

7.1 Introduction

The technical implications surrounding a transition to a renewable energy electrical grid impact all aspects of electricity supply (generation, transmission, distribution, load management). The cumulative impacts will eventually lead to an entirely new system design, a process often referred to as *energy transition, energy revolution*, or *energy system 2.0*.

Changes to the generation sector are manifold. On the one hand, large grid systems are increasingly being fed from generating sources at distant locations such as offshore wind farms or large hydropower from Scandinavia and the Alps. On the other hand, energy grids are also being supplied by a significant increase in smaller, distributed generation systems, particularly photovoltaics (PV). The integration challenges facing solar PV are generation intermittency/variability (solar radiation exposure) and load levels. In order to reduce the requirements for distribution, storage, and additional generation units, electrical grids should be adapted to manage both load and generation levels. That issue is presently treated with high priority by system operators, utilities, and government entities and will be overcome shortly. The focus of this article is on the following three topics: (1) how PV power plants can be adapted to load requirements, (2) possibilities pertaining to load management, and (3) adequate market system design [7].

The article takes the situation in Germany as an example, but most of the results can be applied on any other countries also.

S. Krauter (✉)
Department of Electrical Energy Technology - Sustainable Energy Concepts, University of Paderborn, Pohlweg 55, D-33098 Paderborn, Germany

Photovoltaik Institut Berlin AG, Wrangelstr. 100, D-10997 Berlin, Germany
e-mail: Stefan.Krauter@upb.de; Krauter@PI-Berlin.com

A. Sayigh (ed.), *Photovoltaics for Sustainable Electricity and Buildings*,
DOI 10.1007/978-3-319-39280-6_7

7.2 Measures for PV Generation

On sunny days, with modern-day PV arrays, grid oversupply occurs as a result of the combination of high PV feed and the inflexibility of nuclear as well as fossil fuel power plants in the exiting supply structure. In terms of basic economics, combined oversupply depresses the market price of electricity. Because fixed feed-in tariffs have to be paid, even during periods of oversupply, a considerable amount of money is spent on excess energy.

Figure 7.1 shows the irradiance and the spot-market prices over the course of a sunny day [2]. The system would be more economically balanced with a shift toward more PV power supplied for morning and/or evening power demands, with any corresponding high prices to appear on the spot market. Such a shift could be facilitated via aligning the PV arrays in a more eastern or western orientation—with results as shown in Figs. 7.2 and 7.3.

7.2.1 Adaptation of Location of PV Power Plant

While fossil and nuclear power plants depend on a fuel supply infrastructure plus cooling water intake, the application of PV within Germany (and most other countries) is relatively independent from its location: Throughout Germany, the variation of solar irradiance and corresponding yield is ±15 %.

In order to minimize power transmission costs, it would be useful to install generating capacity as close as possible to the end user. However, from the perspective of the PV investor who receives a fixed feed-in tariff regardless of the location, the costs for land and local workforce often take precedent; therefore, a lot

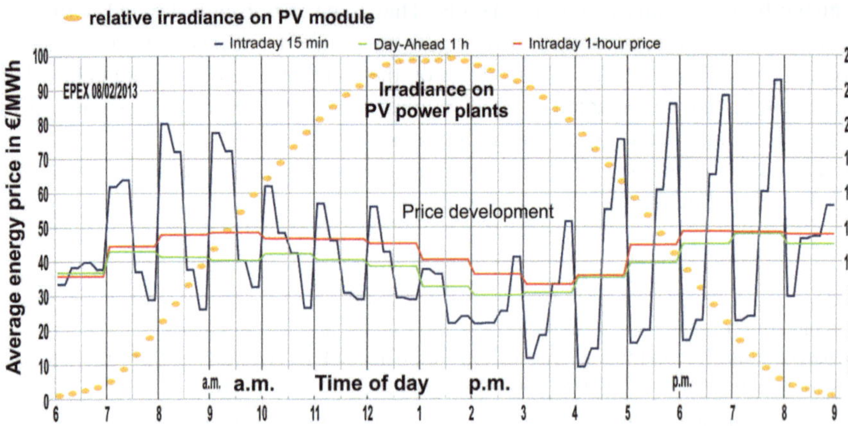

Fig. 7.1 Typical irradiance on a sunny day in Germany and EPEX electricity price development (day-ahead and intraday trading prices [2])

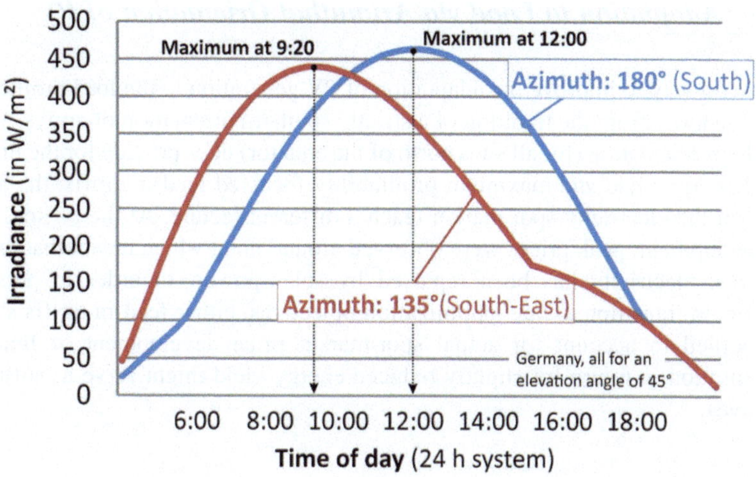

Fig. 7.2 PV generation management via azimuthal deviation from south to 45° southeast: more PV power available during the morning hours (at higher spot-market prices)

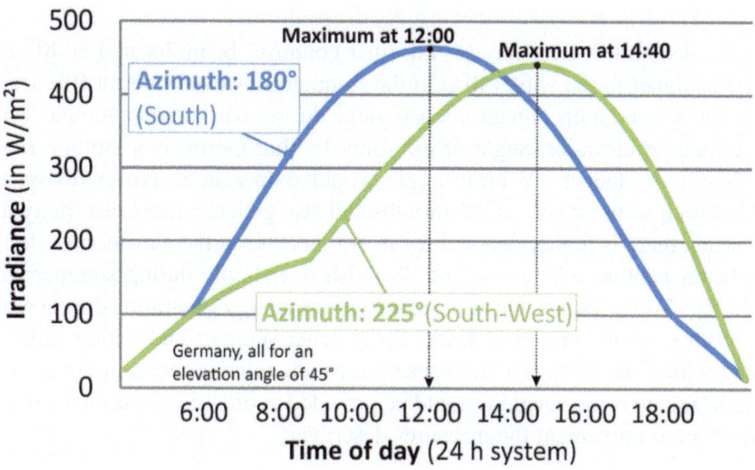

Fig. 7.3 PV generation management via azimuthal deviation from south to 45° southwest: more PV power available during the afternoon hours (at higher spot-market prices)

of PV capacity has been installed far from a majority of consumers in regions with less economic activity and less consumption than more populated regions. As a result, system operators have been forced to invest in additional transmission capacity and grid-balancing measures—all of which ultimately increase the total cost of electricity. Siting PV arrays at former nuclear power plants can often be cost beneficial for both, PV investors and system operators, since they combine excellent transmission infrastructure with at least 1 GW of capacity and cheap land.

7.2.2 Adaptation to Load via Azimuthal Orientation of PV

One useful way to improve the adaptation of PV generation to the load profile is to transition away from the tradition of a strictly southern orientation of arrays. While a southern orientation (for all sites north of the equator) does provide for the highest annual energy yield and maximum profitability for fixed feed-in tariffs, the actual prices on the electricity spot market teach a different lecture. At the beginning of PV development, peak prices were observed around noon which meant that expensive peak electricity has been replaced by solar power. In order to persuade investors to transition to new optimization objectives, either feed-in tariffs should be modified to account for actual spot-market price development or financial compensation in return for slightly reduced energy yield might serve as sufficient incentives.

7.2.3 Adaptation to Seasonal Load Through Modification of PV Array Elevation Angles

Electricity demand in Germany (as all other countries being located at 40°–90° of latitude) is higher in the winter than in the summer; however, present PV arrays are aligned for a maximum annual energy yield during the summer months with an average module elevation angle of 30° only (within Germany's latitude range of 47.5°–54.8°). A steeper PV array angle would reduce large power discrepancies found between summer and winter months and also generate more electricity during winter when there is higher demand for it. An increase in the elevation angle to 60° would better capture winter sun and diminish oversupply during summer. As the graph in Fig. 7.4 represents, the relative increase in energy generation during winter is barely visible, so the energy datasets for average summer and winter months are provided within the table. To compensate for reductions in terms of annual energy harvest, adequate compensation should be provided in order to create an incentive for PV investors to implement the measures described.

7.2.4 PV Tracking Systems

PV tracking systems provide users with the greatest capacity to reduce solar variability and increase power generation throughout the day. Prior to 2008, tracking systems were frequently employed because they helped to generate an additional 30–40 % increase in energy yield. Yet, since 2008, as PV modules have become less expensive, the difference provided by tracking systems, which carry higher installation and operating costs, is not as economically beneficial as they once were. However, with the inclusion of actual spot-market prices and grid system service costs, there could be a renaissance in tracking technologies.

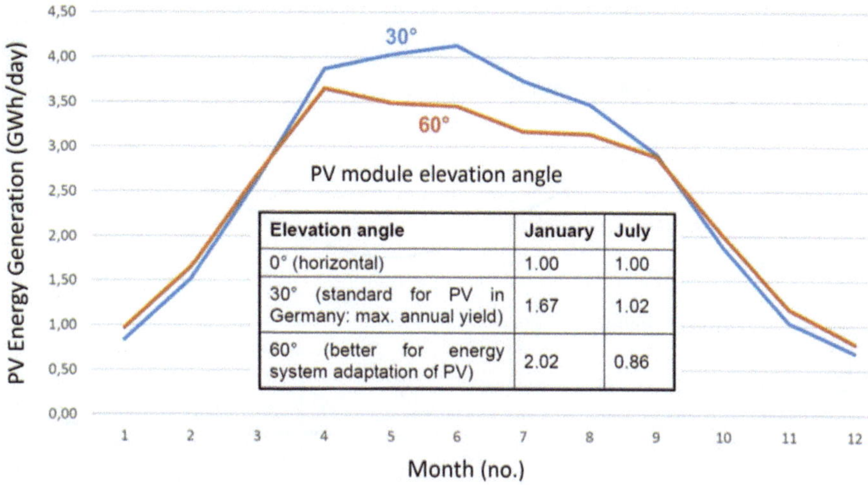

Fig. 7.4 Average daily electrical energy output of a 1 GW PV power plant located in Berlin (latitude, 52.5°N) for 30° and 60° module elevation angles. The included table shows the relative energy output for elevation angles of 0° (ref), 30°, and 60° during the months of January and July

7.2.5 *Thermal Conditioning of PV*

Opportunities for increased efficiency also arise through the use of latent heat storage attached to PV modules. When the warming of PV modules is delayed, PV arrays operate with higher conversion efficiencies and increased power output during the morning hours (see also [3, 4], other cooling strategies [5]). Similar to the tracking systems mentioned in Chap. 7.2.4, the cooling measures carry higher installation costs and are not as economically beneficial as they once were (from the electricity gain point of view). However, with the inclusion of actual spot-market prices and grid system service costs, there could also be a renaissance in cooling technologies.

7.2.6 *Optical Enhancement*

The use of angular-selective/antireflection surface layers or structures can reduce optical reflection, particularly in the morning and in the evening. This is demonstrated for a 90° V-structure on the module surface which leads to increased optical transmission and therefore to higher PV power output (see Fig. 7.5) during periods of high spot-market prices from 9 to 10 a.m. (see Fig. 7.4).

While the PV energy generation gain via V-structures is approximately 3–4 % only [1], that gain is achieved during periods of the day with flat incidence angles and high spot-market prices, so the value of that increase may double (depending on the actual spot-market price variation and its future development). A similar effect

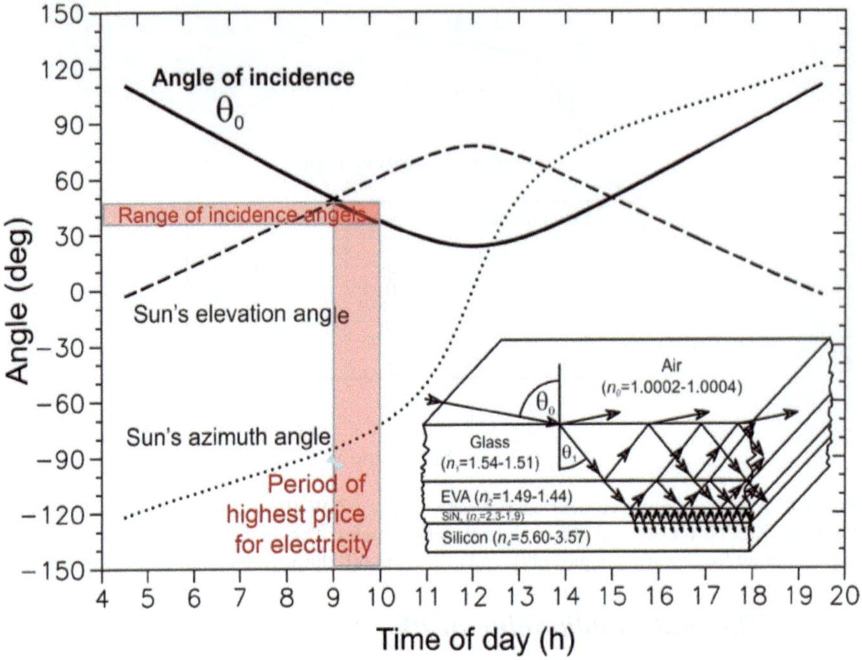

Fig. 7.5 Angle of incidence (*bold line*) during the course of the day; the *red-colored area* represents a period of high spot-market prices and the accordingly relevant angles of incidences. *Inside image* is a cross section of a PV module encapsulation with the optical path of the incidencing irradiance

in optical enhancement plus an additional energy gain from thermal conditioning of up to 10 % can be achieved via a surface water film which has been investigated in [5] (Fig. 7.6).

7.3 Load Management

It is often assumed that the load profile is fixed and constant, with little prospect for customization; however, upon closer look, macro- and microscale opportunities appear for load adaptation. Examples, with a focus on modifications that are relatively easy to implement, are shown below. Possibilities for load adaptation to PV are numerous:

- Flexible tariffs to trigger the use of appliances during favorable periods
- Flexible consumers/producers controlled by grid load, PV output, and predictive use
- Cheap storage located in proximity to end users. Various forms of storage capacity may include:

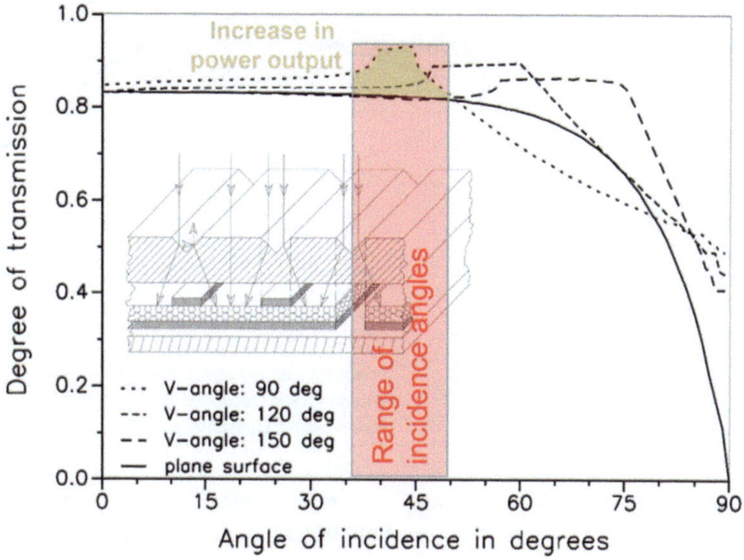

Fig. 7.6 Optical transmission through a typical three-layer module as a function of incidence angle. Plane surface (*bold line*; see also inside image in Fig. 7.5) and different V-structured surfaces (see *inside image* with a V-angle of 90°) together with the relevant incidence angles that correlate to high spot-market prices, as identified in Fig. 7.5. The increase in transmission and power output (*olive color*) is displayed for an advantageous V-angle of 90°

- Heat storage (dishwashers, washing machines)
- Cold storage (refrigerators, freezers, advanced air condtioners)
- Electric vehicles which connect to the grid (G2V, V2G)

7.3.1 *Historical Development of Influence on the Load Curve*

Since the 1960s, load curves have been adapted to an existing or desired generation structure with its characteristic generation profiles. During the very early development of electric power supply, hydropower plants were popular which allow to match demand with generation relatively easy via the control of water flow. Also, this balancing measure operates quite efficient with little loss of energy. Coal and—later—nuclear generation sources, known as *base load power plants*, became increasingly popular around the middle of the twentieth century; however, their flexibility to throttle up/down is limited and requires significant time frames to do so.

Actual electricity requirements and corresponding load profiles in the 1950s and 1960s were characterized by high demand during the daytime and low demand during the nighttime. The most expensive electricity, during peak hours, had to be supplied by additional power plants around midday and in the evening. In order to

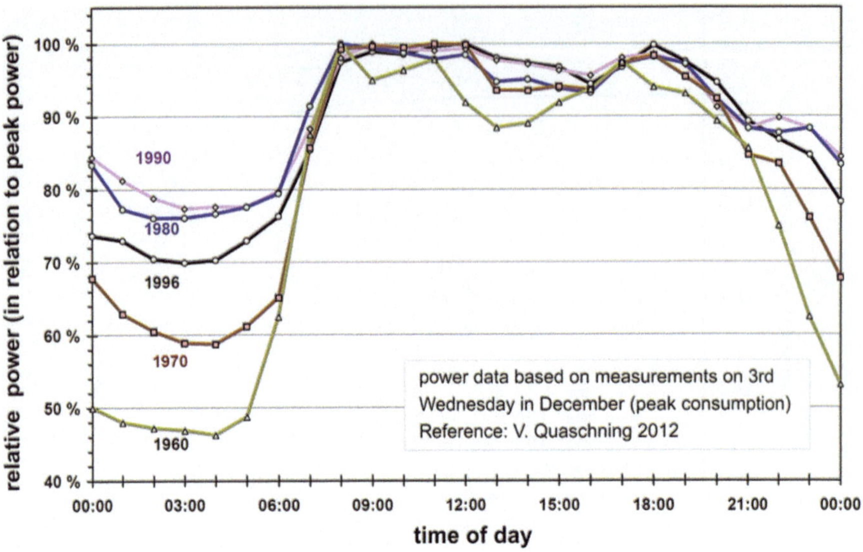

Fig. 7.7 Evolution of the relative diurnal electrical load profile in Germany from the years 1960 until 1996. Until recently, night load has been increased to reduce the day-night variation (load "shaving")

achieve uniform consumption from base load power plants during the day and night (load "shaving"), attractive overnight tariffs have been offered, often in combination with local heat storage options of the consumption units. Some of the most prominent examples are nighttime electric storage heaters and electrical heat pumps. The effect of differentiated overnight tariffs can be observed in Fig. 7.7 with an increase in night consumption being the most apparent. In other words, load curve adjustments are possible and have been successfully achieved in the past. Sometimes the term "load shaping" is used to describe such measures. Due to an increase in solar generating capacity, the load shaping performed over recent decades should be reversed (toward a more pronounced day vs. night cycles—the opposite of load "shaving") in order to better match PV generation with consumption demands and, thus, reduce costs.

7.3.2 *Load Control by "Large Virtual Consumers" (LVC)*

An example of a *large virtual consumer* (LVC) might be end users that employ numerous refrigerators equipped with a latent cold storage and common control mechanisms. That research project investigated the release of *control energy* by *large virtual consumers* for use by the electrical grid. Each of the refrigeration units with latent cold storage is also equipped with a communication device that is able to receive DSM messages via a GSM mobile phone network from an electrical grid

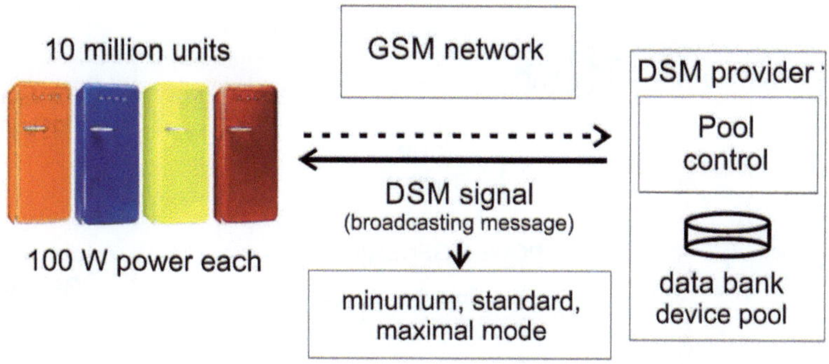

Fig. 7.8 Example of distributed load management: communication structure with DSM signals for influencing load by cooling units (equipped with latent cold storage)

system operator who may ask for a load release (*minimal load*) and load increase (*maximum load*) or allow for device control as necessitated (*standard*) (Fig. 7.8).

DSM was selected to allow for regional control of the broadcasting message, to avoid individual control of each single refrigerator and to maintain privacy. Depending on the operating state of the refrigerator, the command to switch off will be ignored if the products being cooled are at risk; however, due to the condition of latent storage, this happens quite seldom, and, thus, a majority of the refrigerators do participate and effectively contribute to load management.

In Germany, refrigerators and freezers account for 3.6 GW of power which could be theoretically available for load management if they were all equipped with suitable devices. Even if only 10 million new refrigerators were equipped, this would account for 1 GW of load control with high availability at little extra cost.

7.3.3 Power-to-Gas

Power-to-gas (P2G) is a concept in which excess electricity is used to produce hydrogen via electrolysis; then CO_2 is used to convert the hydrogen into synthetic methane. For storage and distribution of methane (and to some extent, hydrogen), existing natural gas infrastructure, such as pipelines, underground storage, and gas stations, can be used. Methane could serve as mobile fuel or as an energy source for heat generation. To produce that gas, excess amounts of electricity from PV may be used. Hydrogen and methane can be converted back into electricity again and serve as an electrical energy storage. Through power-to-gas, the three energy sectors are becoming more integrated and can also support an increase in the proportion of renewable energies for transport and heat generation. Power-to-gas is an effective method to counteract renewable power fluctuations in the network and thus balance generation and load management as well as prevent network congestion. Siting power-to-gas plants close to generation and gas infrastructure can reduce electrical grid loads and other associated costs. Energy transport would be shifted from the

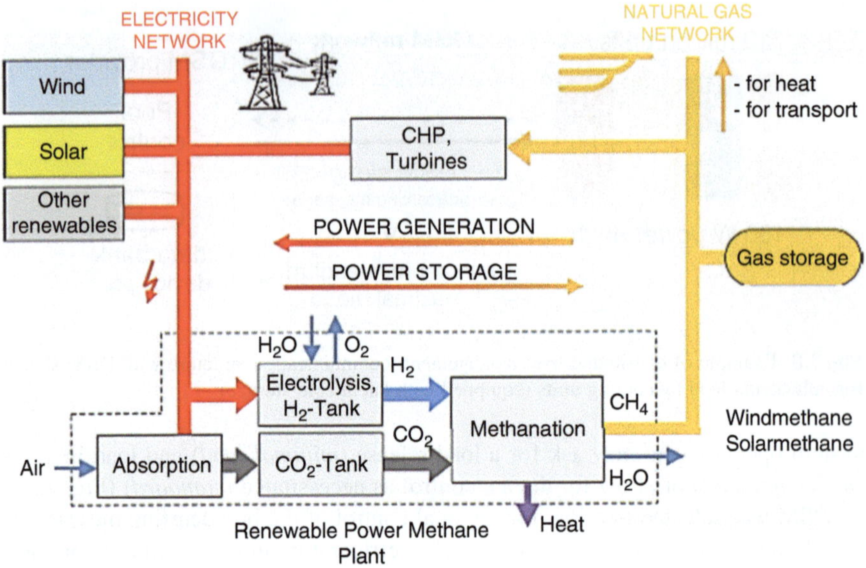

Fig. 7.9 "Power-to-gas" concept [6] using natural gas infrastructure for storage and distribution of energy from PV and other renewable sources

electrical to the gas grid with its higher transmission capacities (e.g., 30 GW for a single pipeline). A model of a power structure operating with 100 % renewable energies using the power-to-gas concept is depicted in Fig. 7.9.

7.3.4 *Electric Mobility to Influence the Load and for Storage*

By 2020 it is estimated that at least 1 million electric vehicles will be operating in Germany. In the private sector, most vehicles are used less than 2 h a day; thus, if a sufficient number of bidirectional power terminals are installed in parking lots, vehicle batteries could be available for load management and grid control services for approximately 20 h a day. Since charging an electric vehicle's battery is completed more quickly during its normal operation than when parked, load adjustments, within limits, can be carried out via adjusting the speed of charge.

Assuming that 70 % of all vehicles have an average storage capacity of 20 kWh and an average charge of 50 %, an additional 7 GWh of capacity could be available to control/balance energy demands. For a further expansion of electric mobility, in a scenario where 90 % of all vehicles in Germany are electric, the available balancing energy would increase to 277 GWh.

Control could be maintained either directly by the user, bundled via *aggregators* (see Fig. 7.10), or via the grid/system operator. It should be noted that utilizing vehicles as storage can be quite expensive, and, thus, load management should

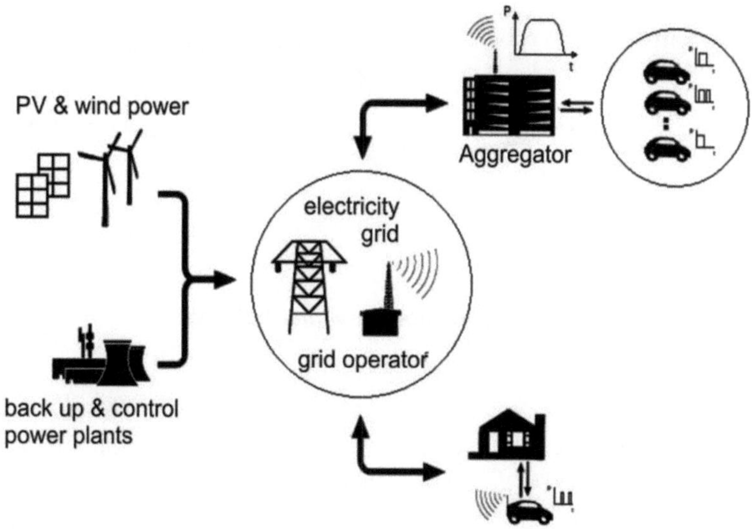

Fig. 7.10 Electric mobility with controlled charge for load management and, in exceptional cases, for intermediate storage

mainly be employed for the charging process—only in emergency cases should vehicle batteries be used as a form of distributed storage.

7.3.5 *Load Management in the Industrial Sector and IT*

Load management can be further employed in industrial sectors (heavy electric grinders and saws) and also for transportation, refrigeration, and heating. Many IT processes such as calculations, data backup, and rendering sites could also serve as forms of load management, but their potential has yet to be sufficiently determined.

7.3.6 *Seasonal Load Adaptation via a Shift in the Holiday Season*

Unfortunately, most electrical energy consumption does not occur in summer (when more PV power is available) but in the winter. While this may be explained, in part, by the demand for more heat and light during the winter, holiday behavior also plays a role. For example, a considerable portion of Germany's economic activity (and of many other countries) slows down during the summer holiday season—this causes a decrease in consumption which leads to overproduction of PV supply. A shift from traditional summer holidays to winter holidays could have

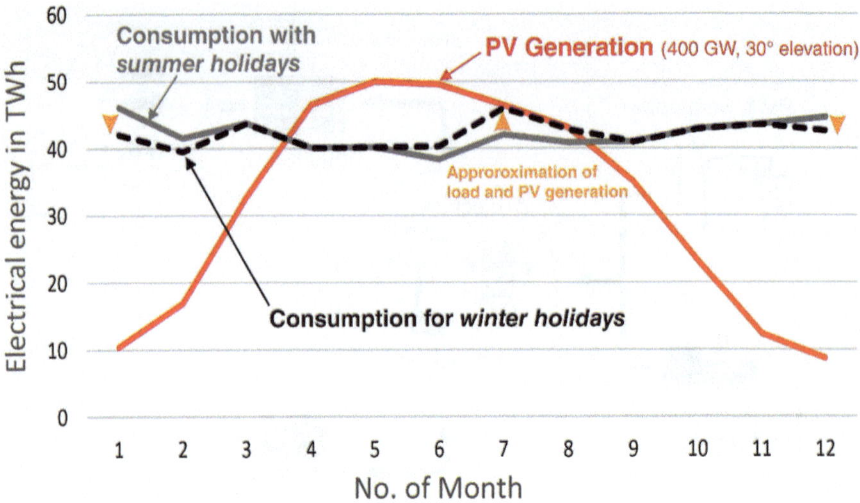

Fig. 7.11 Adaptation of seasonal load curve to the availability of PV power via shifting the summer holidays to winter holidays. Thus, summer consumption is increased, and winter consumption is reduced

a considerable impact on the adaptation of load to PV power availability, as shown in Fig. 7.11. Thus, summer consumption is increased, and winter consumption is reduced, increasing the share of PV in the energy system.

7.4 Marketplace

In order to encourage investment in the concepts shown above, financial incentives are necessary. These incentives can be provided either through government subsidies or through access to the electricity price fluctuations in spot markets. Profitable participation in a spot market requires an open and easily reachable marketplace along with an effective communication infrastructure. Currently, prequalification hurdles are prohibitively high and thus limit participation. An appropriate market design example is shown in Fig. 7.12 and further discussed below.

In this model, participants offer their services (generation, storage, energy management, load shifts) and articulate their needs (power/energy requirements, backup) to a common platform. A deal is made when participants agree on a price. Market regulation and rating systems (e.g., for confidence of delivery), similar to the eBay platform, could be employed between entities. Direct sales from one participant to another via a power purchase agreement (PPA) should be possible. The transport of generated, sold, purchased, or used electricity is becoming more and more simple, accessible, and market based.

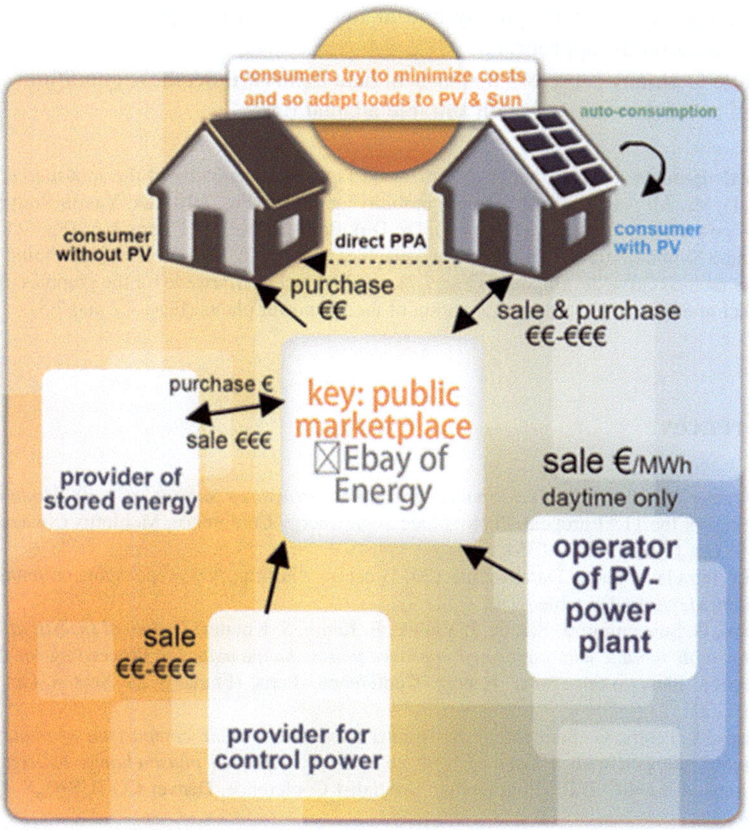

Fig. 7.12 Common market platform through which all participants are enabled to offer and trade their energy services and needs or to perform auto-consumption [7]

In order to fully achieve a PV energy transition, there are administrative and legislative challenges to overcome before an open-market platform is realized, and, without effective market access, many innovative solutions to integrate PV in the energy matrix will not be realized.

7.5 Conclusions

For effective integration of photovoltaics into an energy matrix, a two-way strategy should be employed. One strategy will require different adaptations of the generation process, and the other will require adaptations of the demand. That latter can be achieved via changes in consumer behavior and/or the utilization of appliances and equipment that are capable to carry out load management, e.g., via local storage units. To achieve results quickly and at low costs, a change of behavior on the demand side seems appropriate.

While Germany is used as the basis of data in this article, these conditions can be applied to other countries (with suitable adaptations).

Acknowledgments Sincere thanks to Mr. Ewald Japs for the graphics of the spot-market prices (Fig. 7.1), Mr. Ali Ameli for the "electric mobility" graphics (Fig. 7.10), Dr. Yassin Bouyraaman for the "power-to-gas" graphics (Fig. 7.9), Dr. Dirk Prior for the "LVC" graphics (Fig. 7.8), Mrs. Alessandra Simplicio for the "marketplace" graphics (Fig. 7.12), Prof. Volker Quaschning for the graphics on load curve development (Fig. 7.7), and Mr. Jochen Marwede for the graphics showing the effect of eastern and western orientation of the PV power plants (Figs. 7.2 and 7.3).

References

1. S. Krauter, R. Hanitsch, *Performance of a partly structured surface at a PV-Module*. Proceedings of the 11th European Photovoltaic Solar Energy Conference, Montreux (Switzerland), 12–16 Oct 1992, p. 1351–1354
2. EPEX Intraday Market Data, online (2013) access via http://www.epexspot.com/en/market-data/intraday/intraday-table/-/DE
3. E. Japs, G. Sonnenrein, J. Steube, J. Vrabec, E. Kenig, S. Krauter, *Technical Investigation of a photovoltaic module with integrated improved phase change material*. Proceedings of the 28th European Photovoltaic Solar Energy Conference, Paris (France), 28 Sept–4 Oct 2013, p. 500–502
4. E. Japs, S. Peters, G. Sonnenrein, S. Krauter, *Energy-economic comparison of photovoltaic modules equipped with a layer of conventional and improved phase-change material*. Proceedings of the 40th IEEE Photovoltaic Specialist Conference, Denver CO (USA), 8–13 June 2014
5. S. Krauter, *Increased electrical yield via water flow over the front of photovoltaic panels*. Sol. Energy Mater. Sol. Cells **82**, 137–137 (2004)
6. M. Sterner: *Bioenergy and renewable power methane in integrated 100% renewable energy systems. Limiting global warming by transforming energy systems. Renewable Energies and Energy Efficiency 14*. Kassel University Press. ISBN: 978 3 89958 798 2.
7. S. Krauter, *PV 3.0*, Proceedings of the 27th European Photovoltaic Solar Energy Conference, Frankfurt a.M. (Germany), 24–28 Sept 2012, p. 4520–4522

Chapter 8
Recent Advances in Air-Based Bifacial Photovoltaic Thermal Solar Collectors

K. Sopian, P. Ooshaksaraei, S.H. Zaidi, and M.Y. Othman

8.1 Introduction

In a monofacial solar cells, the front surface is transparent with glass lamination, and the back surface is opaque (Fig. 8.1a). Front surface of the solar cell absorbs the sunlight and converts it into electricity with a metallic grid pattern on the front surface and a blanket metal film on the rear surface. In some field applications, photovoltaic (PV) panels are installed in such a way that rear surface is able to capture diffuse reflected light from ground and other reflective surfaces surrounding the panel. In order to capture this diffuse radiation, researches in the 1980s led to development of a new generation of solar cells able to absorb solar radiation from rear surface [1]. This type of solar cell is referred to as a *bifacial* solar cell. A bifacial solar cell, in contrast with a monofacial solar cell, has identical metallic grids on both front and back surfaces (Fig. 8.1b). Some modifications to conventional monofacial PV panel design are required in order to effectively utilize the bifacial PV panel.

The bifacial solar cell, in contrast with monofacial solar cell, has the capability to absorb solar radiation from the rear surface as well as the front surface, simultaneously. Figure 8.1 compares the mechanism of solar radiation absorption between a monofacial solar cell and a bifacial solar cell. A bifacial module mounted in a highly reflecting environment generates electricity driven from reflected light on the rear aperture as well as illumination on the front side of the device [2].

The transmittance of a monofacial solar cell was measured showing that no light can pass through. This is because at the rear side of this type of cells, there is a thick

K. Sopian (✉) • P. Ooshaksaraei • S.H. Zaidi • M.Y. Othman
Solar Energy Research Institute, Universiti Kebangsaan Malaysia (UKM), Bangi 43600, Selangor, Malaysia
e-mail: ksopian@ukm.edu.my; p_ooshaksaraei@yahoo.com; saleemzaidi@ukm.edu.my; myho@ukm.edu.my

A. Sayigh (ed.), *Photovoltaics for Sustainable Electricity and Buildings*,
DOI 10.1007/978-3-319-39280-6_8

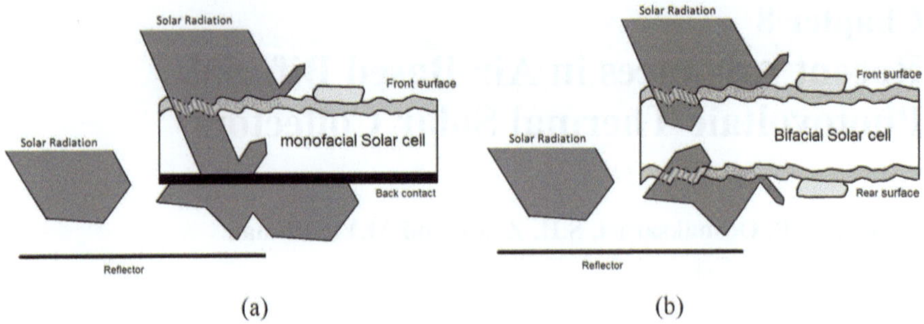

Fig. 8.1 Radiation absorptions of (**a**) monofacial and (**b**) bifacial PV solar cells [2]

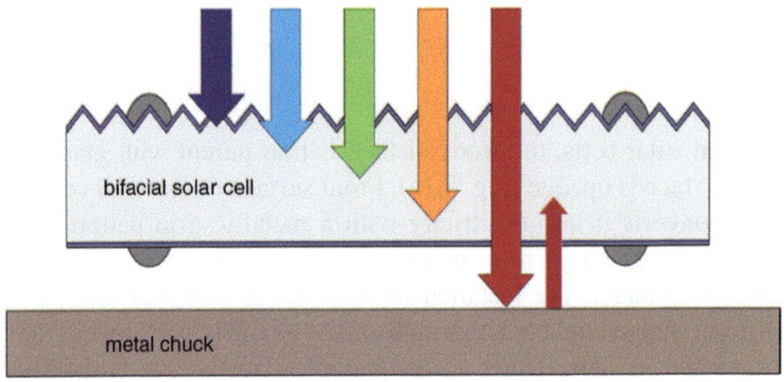

Fig. 8.2 Scheme of light passing though a bifacial solar cell (not to scale) [2]

layer of metal paste, not allowing light crossing through. In the case of bifacial solar cells, the situation is different. Starting from approximately 900 nm (depending on the cell structure), some light passes through the cell (Fig. 8.2) [2].

8.2 Installation Attitude and Reflectors

In conventional bifacial panel applications such as fence integration, front surface usually faces the sun, while the rear surface captures diffuse sky radiation and/or ground reflection based on panel orientation and time of the day. With two absorption surfaces, bifacial solar panels produce more electrical energy than monofacial panels especially in cloudy climates such as Malaysia [3].

Further research on installation attitude bifacial panels emphasizes that a vertically installed (east to west) bifacial photovoltaic solar module setup has two peaks power generation time [4]. One peak is in the morning and the other one happens in the afternoon, while a monofacial PV panel has only one peak power generation through a day, depending on its installation orientation and elevation as shown in Fig. 8.3.

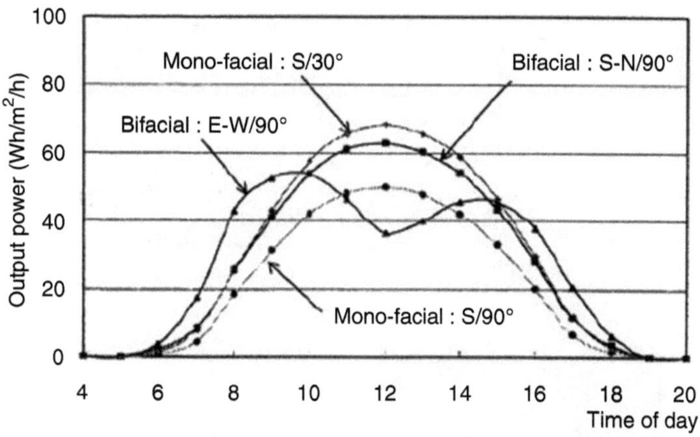

Fig. 8.3 Simulation results on daily generated power distribution of vertically installed bifacial modules compared with monofacial modules [4]

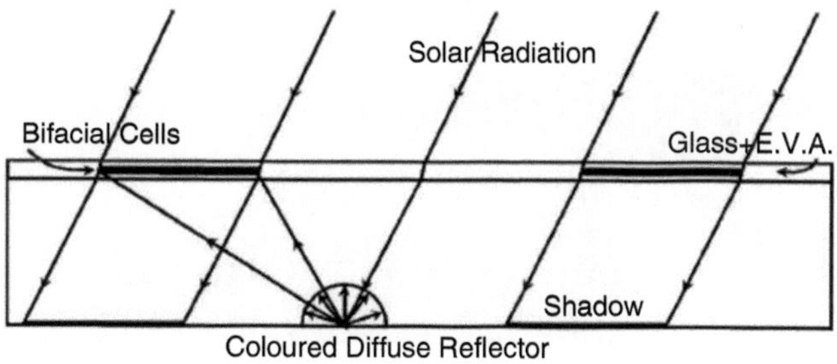

Fig. 8.4 Bifacial PV equipped with diffuse reflector [5]

Though few researches offered vertical installation without backside reflectors, the others spent effort on utilizing various types of reflectors because having a backside absorption is the advantage of bifacial panels in contrast with monofacial ones. This eventually leads to the idea of substituting high-cost photovoltaic material by low-cost reflectors.

Diffuse reflector to return the solar beam radiation to the back surface of bifacial PV cells is shown in Fig. 8.4 [5]. The white color reflector as the best diffuse one with 75 % reflection performance has been incorporated.

Uematsu et al. [6] placed a V-groove reflector with 88 % reflection efficiency at the back, and they left gap between bifacial PV and top glazing to maximize uniform solar gain via multi-reflection by top glazing as shown in Fig. 8.5.

The bifacial panel showed considerable contribution from rear surface. The panel was made of four 5″ × 5″ monocrystalline bifacial cells, a glass to transparent Tedlar. The cells were manufactured by Hitachi Company. Two types of reflectors have been used for indoor characterization work: an aluminum reflector and white-painted aluminum reflectors.

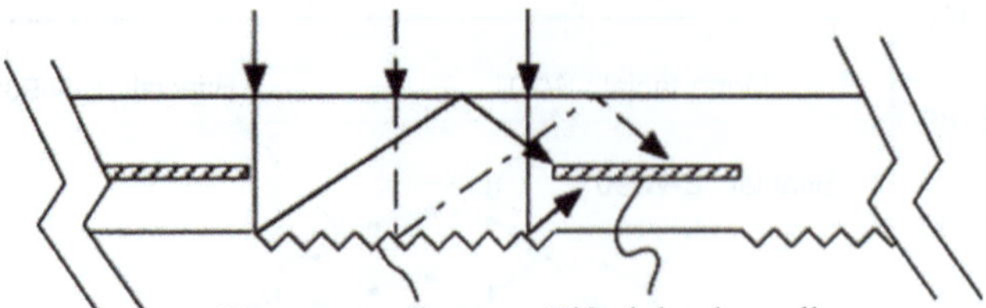

Fig. 8.5 Bifacial equipped with v-groove reflector [6]

Fig. 8.6 Experimental configuration for indoor bifacial solar panel integrated with an external reflector [7]

The bifacial panel outfitted by a semi-mirror type reflector represents the maximum total efficiency at the panel at panel separation 115 mm, while in the case of the diffuse reflector, the maximum efficiency observed at 75 and 115 mm (Fig. 8.6). Effect of the bifacial PV panel separation from reflector on the average electrical power output of the panel is shown in Fig. 8.7, where the output power of the front and the rear apertures are shown separately. The "semi-mirror" and "diffuse" denote the total electrical power generated by the bifacial solar cell equipped with a semi-mirror reflector and a diffuse reflector.

It is possible to reflect the solar radiation inside a glass-to-glass bifacial panel and increase radiation concentration. This interlamination reflector is called "holographic planar concentrator" and shown in Fig. 8.8.

This low-level concentration achieved by ribbons of volume holograms is placed next to photovoltaic cells alternatively. The holographic ribbons increase the panel surface area for fixed number of solar cells but reduce the cost of the module. This cost reduction is achieved by additional electrical energy generation from certain number of solar cells. Although the holograms concentrate addition radiation on existing cells, the temperature reduction effect of the added surface area overcomes

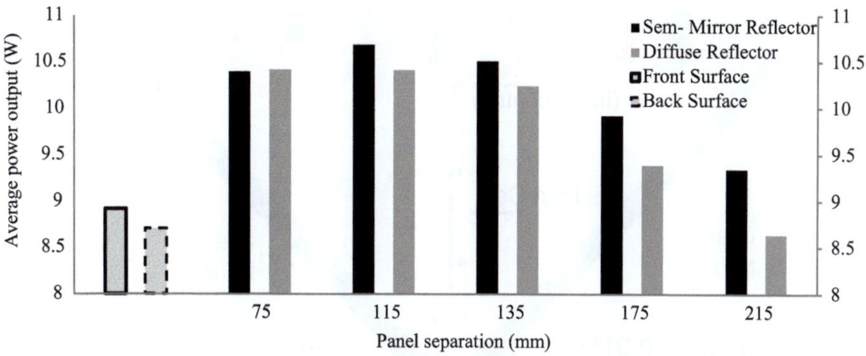

Fig. 8.7 Effect of the bifacial PV panel separation from reflector, on average electrical power output [7]

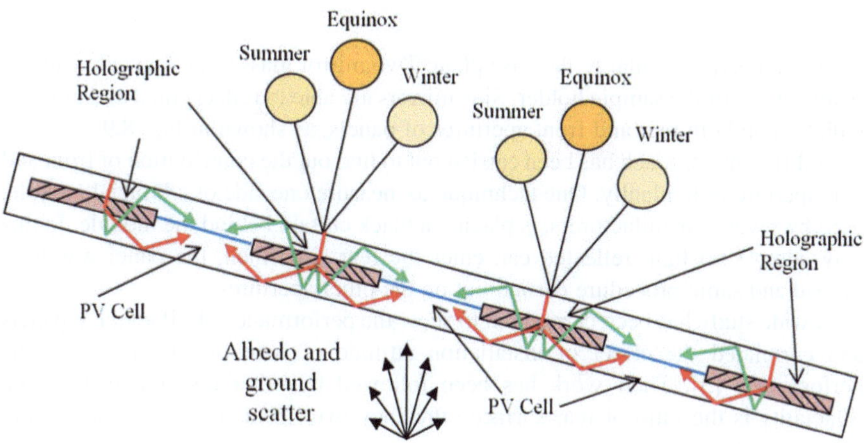

Fig. 8.8 Holographic planar concentrator operating principle with a bifacial silicon cell [8]

the added heat provided by the holograms, which is due to the low concentration and increase in heat transfer surface. The increased heat transfer surface reduces the cell operating temperature and consequently will improve the cell operating efficiency compared to a conventionally assembled panel. The temperature-dependent efficiency phenomenon is due to the negative nature of the P_{max} temperature coefficient for silicon crystalline cells [8].

8.3 Bifaciality

There is no standard to measure performance of bifacial PV modules and also for bifacial solar cells. Ezquer et al. [9] proposed a special setup for I–V characterization of bifacial solar cells. This test method consists of a vertical sample holder

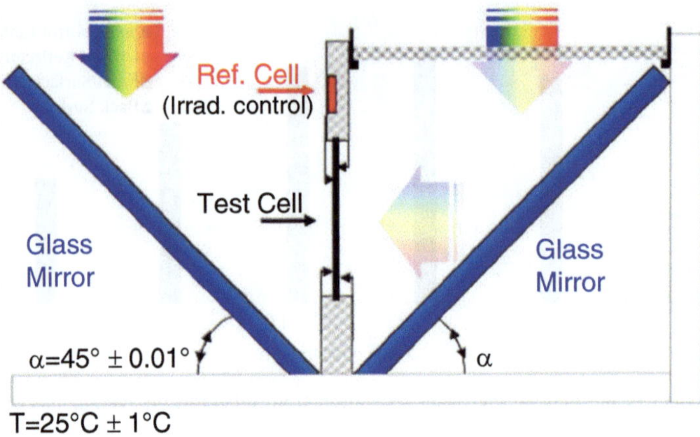

Fig. 8.9 Bifaciality measurement method [9]

positioned perpendicular to the base plate. Two mirror glasses are located symmetrically at 45° of the sample holder. Side mirrors are able to reflect equal and uniform radiation on both rear and front apertures of panels, as shown in Fig. 8.9.

A different approach has been carried out to find out the contribution of front and rear aperture individually. One technique to measure one side of a bifacial module, used by several manufacturers, is placing a black curtain behind the module. In this way, almost no light reflected can enter the rear side; then, the panel was later flipped and same procedure carried out on the other apertures.

A wide study has been carried out to show the performance of bifacial PV panels and correlated the effect of installation attitude of bare bifacial panels to its performance [3]. Their work has been followed by other researchers [6]. The bifaciality is the ratio of rear surface efficiency over front surface efficiency and given as follows:

$$Bifaciality = \frac{Rear\ surface\ efficiency}{Front\ surface\ efficiency}$$

This definition is based on the assumption that same intensity of solar radiation is incident on both rear and front surfaces. Commercially available bifacial solar cells have front and rear efficiencies of 16.6 % and 12.8 %, respectively [10]. Assuming identical microstructures on the front and rear surfaces of solar cells, the lower energy conversion efficiency of the rear surface compared with the front surface may be due to different optical responses of the front and rear surfaces [7, 10]. Electrical energy generated by the rear surface of bifacial solar cells strongly depends on reflection performance of the reflector [11].

Today's bifacial and monofacial modules have nearly the same single side efficiency. Furthermore, the efficiency for a bifacial module is almost equal, no matter from which side it is illuminated [12]. Moreover, the important characteristics were not just the performance (efficiency) but also the bifaciality of panel. Some of the factors affecting the rear contribution can be referred to the illumination conditions such as sun elevation, diffused/global radiation ratio, and reflection of underlying surface or to the module parameters such as bifaciality, module inclination (tilt), distance between rows, and module elevation above underlying surface [11].

The transmittance/absorption factor of the semitransparent and bifacial modules is affected by various parameters, such as the panel orientation, the solar cell packing factor, the efficiency of the solar cells, and the thickness of the modules. However, the influence of solar cell efficiency and PV module's thickness is little [13]. Semitransparent solar cells strongly absorb visible wavelength ($\lambda < 700$ nm) while partially transmits (up to 50 % transmittance) infrared portion of solar radiation (700 nm $< \lambda$) [14].

8.4 Applications of Bifacial PV

When the first photovoltaic cells developed, they were suitable only for space applications, due to their high cost. Later, the use of bifacial solar modules becomes more popular in residential, industrial, and power generation. In addition, there are excellent candidates for applications at low-temperature industrial/agricultural processes besides residential and public applications such as car parks, bus stations, walkway covers, green houses, shelters, balconies, porch coverings, fences, façades, window integrated, and day lighting control [15]. A direct application of bifacial solar panel is albedo collection of the Earth surface, clouds, and the atmosphere. Both their front and rear surfaces capture direct sunlight as well as diffused sunlight from the environment. Here is the point where the use of an appropriate concentrator or reflector can be considered [16]. Moreover, bifacial PV could be invested to residential area such as window integrated, wall integrated, and parking lot integrated.

8.4.1 Wall-Integrated Application

In a wall-integrated application, the panel is installed with an offset distance from the wall. Bifacial solar cells partially cover the panel area. Front surface of bifacial solar cells absorbs a portion of solar radiation, whereas a portion of solar radiation penetrates through transparent vacant space between cells; the wall reflects light back to the rear surface of bifacial cells [17] as shown in Fig. 8.10.

Fig. 8.10 Wall-integrated bifacial PV panel [17]

8.4.2 *Window-Integrated Application*

Window-integrated bifacial PV panel produces electrical energy while permits penetration of faint solar radiation into the interior area for lighting of residential or commercial buildings [14, 15]. Packing factor of solar panel has significant impact on the amount of solar radiation that penetrates the PV module [13]. Space heating and drying applications (partial coverage of drying chamber) could also be considered depending on the climate of the installation site.

A novel multifunctional window shading device, using bifacial silicon solar cells, has been introduced [17] as shown in Fig. 8.11 . The device has achieved 37 % elevated output power compared to monofacial systems. A white semitransparent reflector sheet is placed underneath bifacial panel with separation, which reflects back some of incoming solar radiation on the rear side of cells that incorporate in additional electricity generation. In addition, the light passes through the semitransparent plate into the room behind, depending on the transmittance/absorptance property of reflector.

8.4.3 *Parking Lot Integrated*

Parking-integrated PV panel is an investment on open-area parking lots, where considerable area is available for energy harvesting application. The advantage of

Fig. 8.11 Window-integrated bifacial PV panel [17]

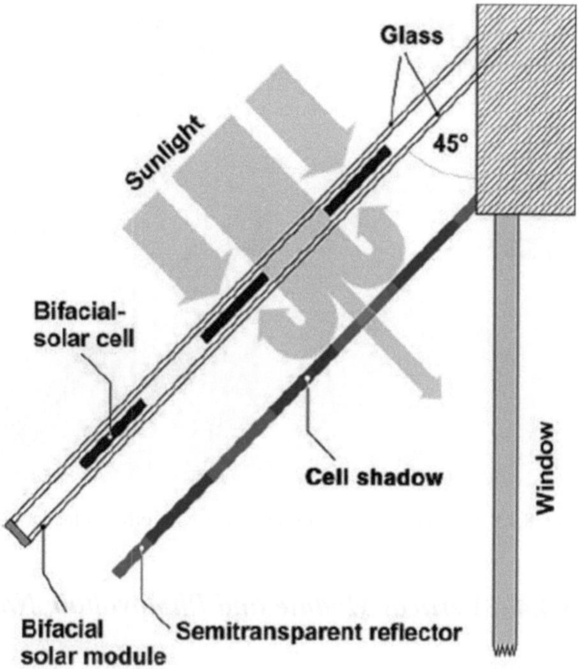

Fig. 8.12 Parking lot-integrated bifacial PV panel [17]

bifacial PV panels in parking lot integrations is that it provides shading for vehicles same as monofacial panels, but the bifacial panels also absorb reflected sunshine from backside reflector as shown in Fig. 8.12 .

Fig. 8.13 Bifacial Zurich–Aubrugg plant with bifacial technology [18]

8.4.4 Vertical Module and Photovoltaic Noise Barriers

The idea of using PV panels as noise barrier was introduced in Switzerland, a country with high installed PV power capacity [2]. According to Nordmann et al. [18], the photovoltaic noise barriers along motorways and railways are among the most economic applications of grid-connected PV with the additional benefits of large-scale plants and with no extra land consumption.

Figure 8.13 shows the world's first PV noise barrier bifacial plant, located near Zurich, on a north–south motorway flyover [18]. The bifacial modules act as sound reflecting element, while one side is exposed to the morning sun and the other to afternoon sun. The back surface had a slightly lower efficiency (bifaciality factor less than one). Hence, the annual performance was therefore lower than might be expected for a fully symmetric bifacial array. Theoretically, the annual yield should be equal to or higher than the yield of a south-facing array.

Sound barriers and vertical installation of bifacial solar systems presented an innovative application for this type of devices. The increased power output in this case will depend on their spatial installation. The results will always be higher than monofacial systems.

8.5 Bifacial PVT Panels

Photovoltaic thermal (PVT) collectors are especially designed to generate both electrical and thermal energies simultaneously. A solar hybrid panel generates more energy per unit area compared to ordinary photovoltaic (PV) or thermal

(T) collector separately side by side with same total collector area [19]. Temperature increase of the PV module is the main reason for reduction of the electrical efficiency. The other two main reasons are the packing factor (PF) of PV module and ohmic losses between two consecutive solar cells. The overall electrical efficiency of the PV module can be increased by increasing the packing factor (PF) and reducing the temperature of the PV module by using the thermal energy associated with the PV module [20, 21]. A typical residential PVT collector consists of a flat plate PV panel, attached on the top of the metallic absorber plate. The PV cells are composed of semiconductor materials that convert high-energy photons of incident solar radiation into electricity. The lower energy photons are absorbed by the PV panel and generate heat within the cells. This heat is rejected to the ambient conditions while also being removed from behind the PV cells by a heat transfer fluid. The extracted heat from the PV panels offers multiple benefits. The heat is used for thermal loads while also improving the PV electrical efficiency due to its lower operating temperature. Both the amount of thermal energy and its quality are important factors when considering the value of the output heat. The performance of a PVT system must be characterized by both the amount of electrical energy and useful heat produced.

8.5.1 Bifacial PVT with Reflectors

A bifacial PVT panel with reflectors is shown in Fig. 8.14 . The experimental model of the PVT bifacial PV module was constructed and studied, and the overall solar energy utilization efficiency for the system was about 60 % with

Fig. 8.14 Bifacial PVT solar collector [22]

an electric efficiency of 16.4 %. Moreover, the bifacial PVT panel equipped with the aluminum reflectors generates 40 % additional electrical and thermal energies [22].

8.5.2 Multi-pass Bifacial PVT

There are many different designs that are used for the heat removal in PV/T collectors. Most notably, the collector can use either air or an aqueous solution as its heat removal fluid. But most of the existing PVT panel designs are inappropriate for bifacial solar cells because the absorber plate covers the rear surface of bifacial solar cells [23–26]. The absence of solar radiation on the rear surface of bifacial solar cells destroys the benefit of dual surface solar radiation absorption of bifacial solar cells. Panel modification is an essential action; the absorber plate should be substituted by a reflector to reflect back solar radiation to the rear surface of bifacial solar cells. Figure 8.15 represents four air-based multi-pass bifacial PVT

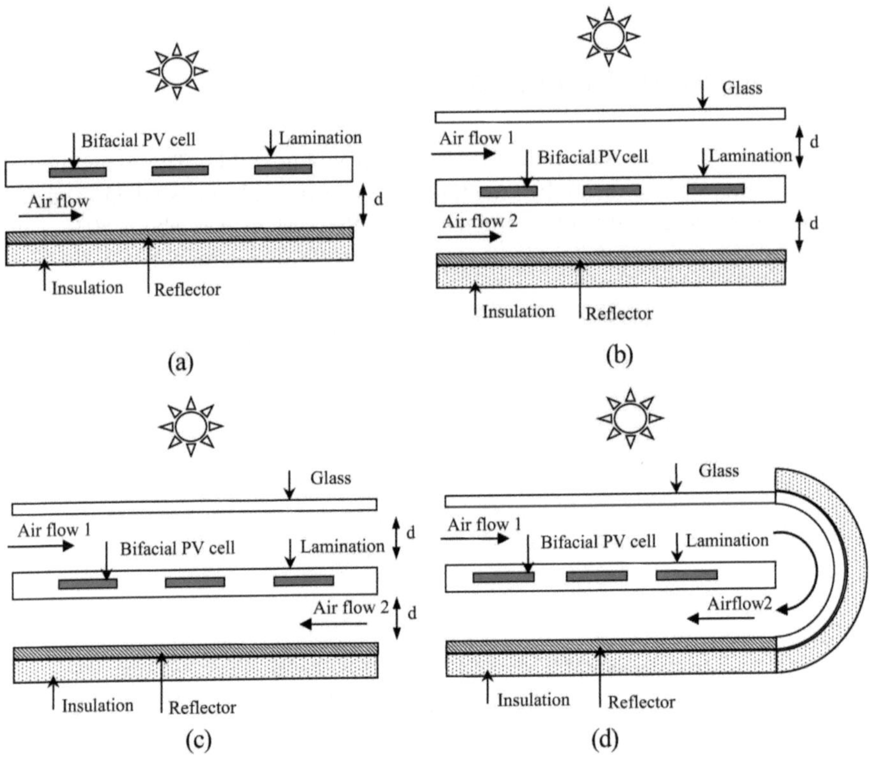

Fig. 8.15 Cross section view of (**a**) Model 1: single-path, bifacial PVT panel. (**b**) Model 2 double-path, parallel flow, bifacial PVT panel. (**c**) Model 3: double-path, counterflow, bifacial PVT panel. (**d**) Model 4: double-pass, returning flow, bifacial PVT panel [26]

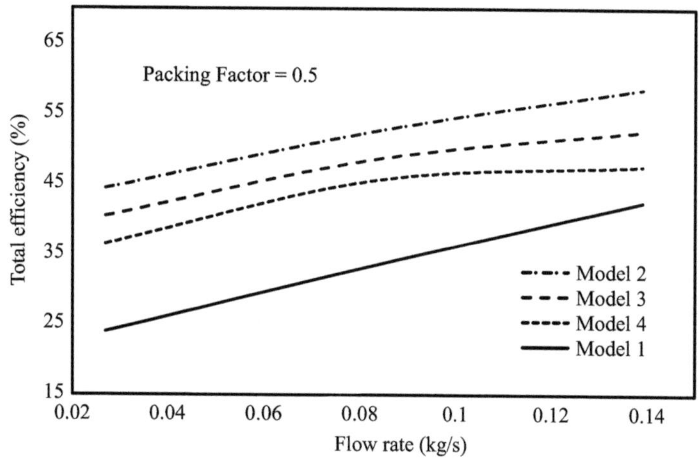

Fig. 8.16 Total efficiency with different flow rates [26]

panels developed based on the existing monofacial PVT panel designs and the requirements of bifacial solar cells.

Figure 8.16 shows the total efficiency of the bifacial PVT panel with different flow rates [26]. Panel model 2 has the highest energy efficiency, followed by models 3, 4, and 1. In addition, Panel model 1 has the highest exergy efficiency. But panel models 2, 3, and 4 have a quiet similar performance from the exergy point of view. Panel model 1 is recommended for applications where higher electrical output is required. Panel mode 2 is recommended for applications where higher thermal output is required. All four designs have a higher performance at higher mass flow rate and also higher packing factor with respect to both the first and second laws of thermodynamics. Based on the cost-effectiveness analysis, the optimum air flow rate is 0.06–0.08 kg/s from energy point of view.

Under such economically favorable conditions, the use of photovoltaic thermal collectors became ideal for a wide variety of applications. But economic analysis is crucial in any development procedure. The design should be feasible and viable. Here we present the annual cost (AC) and the annual energy gain (AEG) of the bifacial PVT collectors. The ratio of AC/AEG or the cost–benefit ratio is presented to make it feasible for the user to select optimum design features which correspond to minimum AC/AEG. The main objective of the optimization study is to be able to specify design parameters, a combination of which can extract maximum possible thermal and electrical energy out of the available solar energy. If photovoltaic thermal solar collectors are to be economically beneficial, they must have the capability to collect maximum amount of solar energy at minimum possible cost.

Figures 8.17 and 8.18 show the cost–benefit ratio or the AC/AEG as a function of the flow rate for different packing factors (PF) of bifacial PV. Packing factor is the area of the of the collector area covered by the bifacial solar cell. It is expected that

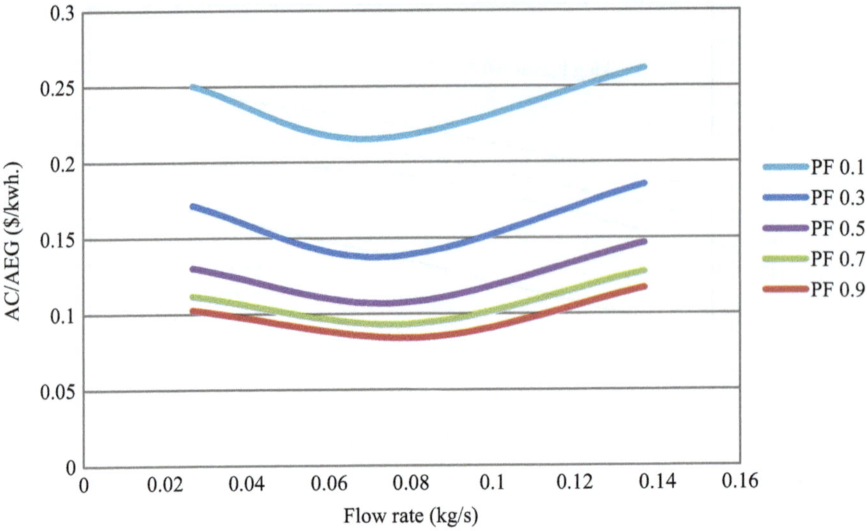

Fig. 8.17 AC/AEG as a function of the mass flow rate at different packing factors (panel model 1) [26]

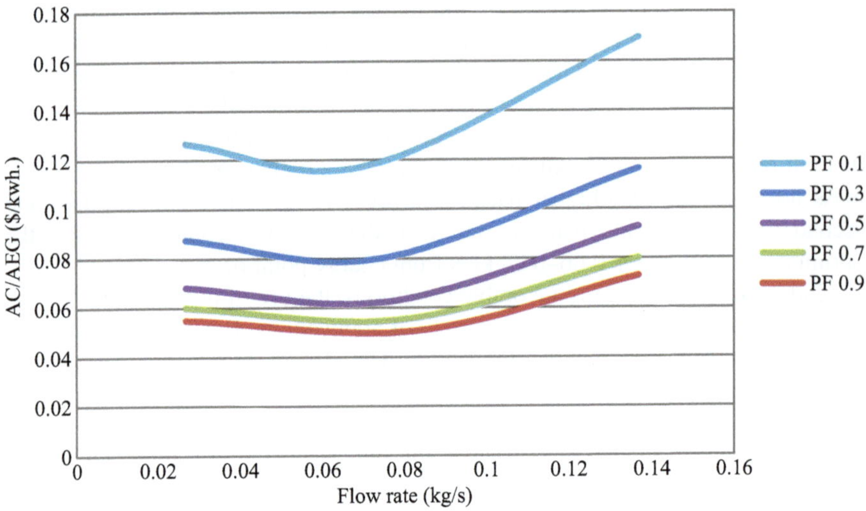

Fig. 8.18 AC/AEG as a function of the mass flow rate at different packing factors (panel model 2) [26]

panels with higher packing factor achieve the lowest AC/AEG at relatively higher mass flow rate which is due to higher radiation absorption in PV panel and higher operation temperature consequently. The higher flow rate contributes to lower operation temperature and consequently higher PV panel electrical efficiency.

8.6 Conclusions

The advantage of bifacial PVT especially in flat plate and low concentration collectors are that the increase in electrical energy production with little increase in investment cost. Bifacial PVT panels offer a wide variety of applications especially in building integrated photovoltaic components. Bifacial PVT produces energy (electricity) from exterior walls, windows, open-area parking lots, and roof. Bifacial PVT provides shading and faint light penetration. Bifacial PVT could also be considered as an architecture and HVAC components.

Research and development on static flat plate bifacial panels will result to higher market share than the conventional monofacial panels because of the additional electricity production from the rear surface. Optical enhancement devices such as external reflectors and compound parabolic collectors must be explored since high packing factor means less solar radiation absorption by back surface of solar cells. Other optical enhancement devices such as Fresnel lens can be used. It is possible to replace to top glass cover of dual duct panels with a Fresnel lens to concentrate light on front surface and back surface of bifacial PV cells. The present uses mostly the monocrystalline silicon type photovoltaic cell. It is recommended carrying out a research on the other types of bifacial cells to find out the optimum bifacial cells for PVT application such as dye-sensitized solar cells and other thin films such as CIGS and CdTe.

References

1. A. Luque, A. Cuevas, J.M. Ruiz, Double-sided n + -p-n + solar cell for bifacial concentration. Sol. Cells **2**(2), 151–166 (1980)
2. C. Duran, Bifacial solar cells: high efficiency design, characterization, modules and applications. Doctoral Dissertation, Department of Physics, Universität Konstanz, Germany (2012)
3. T. Joge, Y. Eguchi, Y. Imazu, I. Araki, T. Uematsu, K. Matsukuma, Applications and field tests of bifacial solar modules. Photovoltaic Specialists Conference, 2002. Conference Record of the Twenty-Ninth IEEE, hlm (2002), pp. 1549–1552
4. T. Joge, Y. Eguchi, Y. Imazu, I. Araki, T. Uematsu, K. Matsukuma, Basic application technologies of bifacial photovoltaic solar modules. Electr. Eng. Jpn **149**(3), 32–42 (2004)
5. A. Moehlecke, I. Zanesco, A.C. Pan, T.C. Severo, A.P. Mallmann, Photovoltaic module with coloured diffuse reflectors. Proceedings of the European Photovoltaic Solar Energy Conference, Munich, Germany (2001), pp. 785–787
6. T. Uematsu, Y. Yazawa, K. Tsutsui, Y. Miyamura, H. Ohtsuka, T. Warabisako, T. Joge, Design and characterization of fat-plate static-concentrator photovoltaic modules. Sol. Energy Mater. Sol. Cells **67**, 441–448 (2001)
7. P. Ooshaksaraei, K. Sopian, R. Zulkifli, M.A. Alghoul, S.H. Zaidi, 2013. Characterization of a bifacial photovoltaic panel integrated with external diffuse and semimirror type reflectors. Int. J. Photoenergy **2013**, Article ID 978234, 10 pages (2013)
8. J.E. Castillo, J.M. Russo, R.K. Kostuk, G.A. Rosenberg, Thermal effects of the extended holographic regions for holographic planar concentrator. J. Photonics Energy **1**(1), 015504-015504-015513 (2011)

9. M. Ezquer, I. Petrina, J.M. Cuadra, A. Lagunas, Design of a special set-up for the I-V characterization of bifacial photovoltaic solar cells. Proceedings of the 23rd European Photovoltaic Solar Energy Conference and Exhibition, 1–5 September 2008, Valencia, Spain (2008), pp. 1553–1556
10. L. Yang, Q.H. Ye, A. Ebong, W.T. Song, G.J. Zhang, J.X. Wang, Y. Ma, High efficiency screen printed bifacial solar cells on monocrystalline CZ silicon. Prog. Photovolt. Res. Appl. **19**(3), 275–279 (2011)
11. C. Duran, H. Deuser, R. Harney, T. Buck, Approaches to an improved IV and QE characterization of bifacial silicon solar cells and the prediction of their module performance. Energy Procedia **8**, 88–93 (2011)
12. S. Guo, T.M. Walsh, M. Peters, Vertically mounted bifacial photovoltaic modules: a global analysis. Energy **61**, 447–454 (2013)
13. T.Y.Y. Fung, H. Yang, Study on thermal performance of semi-transparent building-integrated photovoltaic glazings. Energy Build. **40**(3), 341–350 (2008)
14. S. Yoon, S. Tak, J. Kim, Y. Jun, K. Kang, J. Park, Application of transparent dye-sensitized solar cells to building integrated photovoltaic systems. Build. Environ. **46**(10), 1899–1904 (2011)
15. P. Ooshaksaraei, R. Zulkifli, S.H. Zaidi, M.A. Alghoul, A. Zaharim, K. Sopian, Terrestrial applications of bifacial photovoltaic solar panels. Proceedings of the 10th WSEAS International Conference on System Science and Simulation in Engineering, Penang, Malaysia (2011), pp. 128–131
16. A. Cuevas, A. Luque, J. Eguren, J. del Alamo, 50 per cent more output power from an albedo-collecting flat panel using bifacial solar cells. Sol. Energy **29**(5), 419–420 (1982)
17. R. Hezel, Novel applications of bifacial solar cells. Prog. Photovolt. Res. Appl. **11**(8), 549–556 (2003)
18. T. Nordmann, A. Froelich, A. Goetzberger, G. Kleiss, G. Hille, C. Reise, E. Wiemken, V. van Dijk, J. Betcke, N. Pearsall, The potential of PV noise barrier technology in Europe. Proceedings of the 16th European Photovoltaic Solar Energy Conference and Exhibition, Glasgow, Scotland (2000), pp. 1–5
19. S. Agrawal, G.N. Tiwari, H.D. Pandey, Indoor experimental analysis of glazed hybrid photovoltaic thermal tiles air collector connected in series. Energy Build. **53**, 145–151 (2012)
20. T.T. Chow, Performance analysis of photovoltaic-thermal collector by explicit dynamic model. Sol. Energy **75**(2), 143–152 (2003)
21. H.A. Zondag, D.W. de Vries, W.G.J. van Helden, R.J.C. van Zolingen, A.A. van Steenhoven, The yield of different combined PV-thermal collector designs. Sol. Energy **74**(3), 253–269 (2003)
22. B. Robles-Ocampo, E. Ruíz-Vasquez, H. Canseco-Sánchez, R.C. Cornejo-Meza, G. Trápaga-Martínez, F.J. García-Rodriguez, J. González-Hernández, Y.V. Vorobiev, Photovoltaic/thermal solar hybrid system with bifacial PV module and transparent plane collector. Sol. Energy Mater. Sol. Cells **91**(20), 1966–1971 (2007)
23. T.T. Chow, A review on photovoltaic/thermal hybrid solar technology. Appl. Energy **87**(2), 365–379 (2009)
24. M.A. Hasan, K. Sumathy, Photovoltaic thermal module concepts and their performance analysis: a review. Renew. Sustain. Energy Rev. **14**(7), 1845–1859 (2010)
25. R. Kumar, M.A. Rosen, A critical review of photovoltaic–thermal solar collectors for air heating. Appl. Energy **88**(11), 3603–3614 (2011)
26. P. Ooshaksaraei, K. Aghili, K. Sopian, R. Zulkifli, S. Zaidi, Steady state characterization of bifacial solar cells at different configurations of air-based photovoltaic thermal solar panels. J. Renew. Sustain. Energy **6**(3), 033140 (2014)

Chapter 9
30 Years of Living in a Solar House in Colorado

Keith Emery and Pat Emery

Having met in 1980 while working at the National Renewable Energy Laboratory, we were inspired by the vision of a renewable energy future to raise our family in a passive solar home. A basic, factory-cut, trilevel house design was selected in 1985 and intentionally sited and modified according to passive solar design principles to optimize conservation and solar gain, with a plan to add photovoltaics as soon as feasible (Fig. 9.1).

As general contractors, we were responsible for the foundation, painting, insulation, and interior work, including wiring, plumbing, and heating. The poured cement foundation is insulated with Styrofoam to R20 for 61 cm and R10 for another 61 cm down to the footer. The vaulted ceiling above the main level is insulated with fiberglass to R38 and the roof above the upstairs bedrooms to R57. The walls contain the maximum possible R19.

The main level is on the south side of the house with living room, kitchen, and sunroom facing the backyard and with the entry to the house on the west end. The two-story north side has two bedrooms and bath upstairs and a basement with a finished bedroom, bath, den, and laundry area. The main level and upstairs hallway are open under a vaulted ceiling that lets air flow unimpeded from the sunroom (10'x 36') through the main room, to the basement and upstairs (ea. @ $15' \times 36'$). An operable roof window in the north side bathroom is located as high as possible on the roof for ventilation and to facilitate air circulation throughout the house.

The floors on the main room and sunroom are 6″ thick cement slabs. Slate tile flooring transmits the heat of the sunlight to the cement thermal mass. The eaves were designed to prevent direct light from entering the sunroom during summer. Thermal management of the house is most difficult on warm spring and fall days, when the sun is not high enough to be blocked by the eves. Lowering curtains fitted

K. Emery (✉) • P. Emery
NREL, 15013 Denver West Parkway, Golden, CO 80401, USA
e-mail: keith.emery@nrel.gov

A. Sayigh (ed.), *Photovoltaics for Sustainable Electricity and Buildings*,
DOI 10.1007/978-3-319-39280-6_9

Fig. 9.1 South side of the house

for the argon-filled double pane windows in the sunroom reduces this heat load, as well as reducing heat loss on cold winter nights.

In 2007, a SunPower 7.6 kW PV system was installed on the roof (36 modules that are 1.24 m^2 and 201 W each). The manufacturer SunPower was chosen because they are vertically integrated, approve all components, and certify the installer. We chose Boulder, CO-based Namasté Solar, an installer with a long track record of installing systems with no problems, as we knew poor workmanship can lead to roof leaks, ground faults or other damage that may not be discovered for years. As part of a roof replacement after damage during a recent windstorm, our insurance paid $4868 for taking down the system and putting it back up without voiding its warranty. This is about 25 % of the net price we paid for the installed system after the significant rebates and incentives we received.

Our PV system has produced 50.0 MWh since its installation. The PVWatts model [1] predicted an ac energy of 11 MWh annually or 49.5 MWh over the lifetime to date of the system (Fig. 9.2). The several years of solar energy data, measured 11 miles from the house [2], show that 9.93 MWh/m^2 of solar energy fell on the system during that time, giving a net efficiency of 11% for the entire system, including inverter losses.

The system powers a heat pump for cooling and for heating on all but the coldest days. A natural gas furnace forced air system acts as backup, needed only on cold early mornings or after several cloudy days. Natural gas also heats water in a

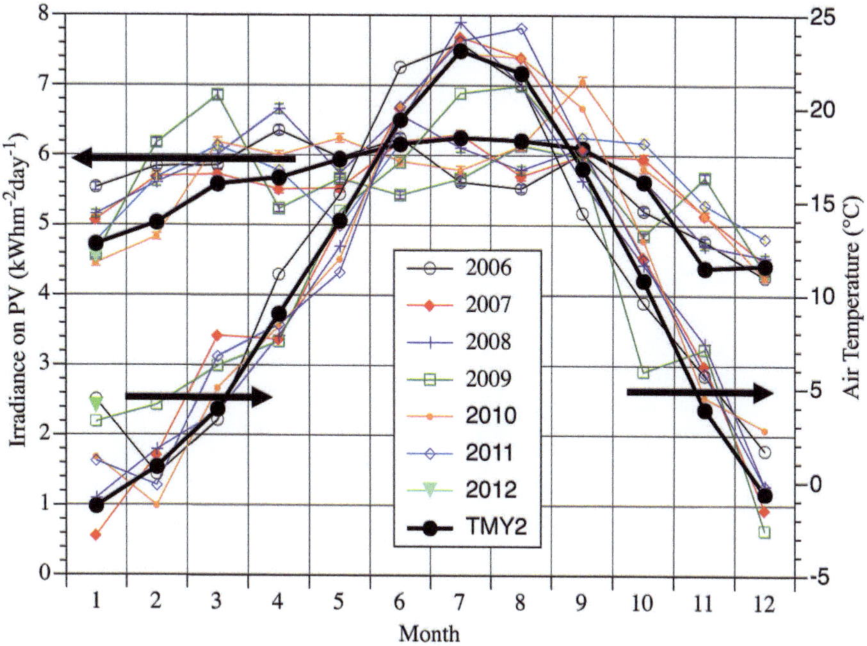

Fig. 9.2 Typical meteorological year in PVWatts and measured data from SRRL [2]

tankless on-demand system. These are highly efficient systems, and their combination has made solar hot water panels financially unwarranted. Our energy bill averages $25 per month, including fees and taxes. About half the total is returned to us annually as payment for power we provide to the grid.

Over the years we have enjoyed living in an energy-efficient, solar home. The daily and seasonal tasks associated (curtains, fans, etc.) were easily integrated into our lives. We have been spared the anxiety of constantly rising energy prices. The excellent thermal behavior of the house has been comfortable; the bright daylighting has allowed for healthy houseplants and a cheery atmosphere.

References

1. PVWatts website, www.nrel.gov/rredc/pvwatts/
2. NREL Solar Radiation Research Laboratory Baseline measurement system, website http://www.nrel.gov/midc/srrl_bms/

Conclusions

In this book the authors have looked at PV from a variety of viewpoints. It can be seen that the overriding vision is for PV to become a major global supplier of electricity in the twenty-first century at affordable cost cheaper than fossil fuels within the context of potentially catastrophic climate change, which was largely triggered by the industrial revolution and the use of fossil fuels, emphasizing the ease of use and installation and its almost pollution-free generation.

Examples in the book look at both macro megawatt power stations and micro stand-alone generation systems, and an historical overview is given. One of the remarkable achievements in the development of PV has been the considerable improvement in efficiency accompanied by greatly reduced costs. In 1977 the cost of one Watt of electricity produced by PV was US$76.67 which has fallen to US$0.36 per Watt today. While in 1960 the efficiency of a crystalline silicon cell was 6 % and today is more than 25 %.

Because one of the major applications for PV panels is on roofs and facades, PV should also be considered as a building material in itself; architects and the building industry are urged to integrate PV into their designs rather than making PV a haphazard add-on.

As an example of the use of PV globally in 2014 crystalline silicon cells continue to have a larger market share compared to that of thin films: 43.1 GW of crystalline silicon cells while thin films were 4.4 GW. In the same year Germany, China, Japan, North America, and Italy continue to dominate the field in PV installation. In terms of efficiency under normal operating lab scale test conditions, most types of cell have exceeded 21 %, while amorphous silicon has reached over 13 %.

The history of PV is considered as well as looking at one exemplar of a long-term stand-alone solar house. On the other hand power plant layout is discussed taking into consideration load requirement and management of load shifting, load adoption, and appropriate market design.

New concepts in cell efficiency and production from hybrid photovoltaic-thermal cells (PV/T) and bifocal cells are described and it is envisaged they will produce up to five times the voltage of normal cells and hence more energy.

A. Sayigh (ed.), *Photovoltaics for Sustainable Electricity and Buildings*,
DOI 10.1007/978-3-319-39280-6

A full analysis is given of solar radiation applications for PV use as well as concise notes for teaching the mathematical concepts associated with PV theory and practice.

The future of PV continues to be one of intense research and development where thin films will overtake crystalline silicon in cost and efficiency. It is foreseen that the role of PV in the generation of electricity will overtake other forms of renewable energy and that 50 % of all new buildings will incorporate PV in their design while the debate over the use of nuclear fuel continues despite doubts as to its long-term economic and environmental viability.

Index

A. Sayigh (ed.), *Photovoltaics for Sustainable Electricity and Buildings*,
DOI 10.1007/978-3-319-39280-6

MIX
Papier aus verantwortungsvollen Quellen
Paper from responsible sources
FSC® C105338

If you have any concerns about our products,
you can contact us on
ProductSafety@springernature.com

In case Publisher is established outside the EU,
the EU authorized representative is:
Springer Nature Customer Service Center GmbH
Europaplatz 3, 69115 Heidelberg, Germany

Printed by Libri Plureos GmbH
in Hamburg, Germany